The Best Breeds of British Livestock
A Practical Guide For Farmers and Owners of Livestock in England

by John Watson F.L.S.

with an introduction by Jackson Chambers

This work contains material that was originally published in 1898.

This publication is within the Public Domain.

This edition is reprinted for educational purposes
and in accordance with all applicable Federal Laws.

Introduction Copyright 2018 by Jackson Chambers

Self Reliance Books

Get more historic titles on animal and stock breeding, gardening and old fashioned skills by visiting us at:

http://selfreliancebooks.blogspot.com/

Introduction

I am pleased to present another title in the "Cattle" series.

The work is in the Public Domain and is re-printed here in accordance with Federal Laws.

As with all reprinted books of this age that are intended to perfectly reproduce the original edition, considerable pains and effort had to be undertaken to correct fading and sometimes outright damage to existing proofs of this title. At times, this task is quite monumental, requiring an almost total "rebuilding" of some pages from digital proofs of multiple copies. Despite this, imperfections still sometimes exist in the final proof and may detract from the visual appearance of the text.

I hope you enjoy reading this book as much as I enjoyed making it available to readers again.

Jackson Chambers

CHAMPION DEXTER KERRY COW "ROSEMARY."

PREFATORY NOTE

IN spite of the various vicissitudes of our home Agriculture, stock-keeping is still its great prop; and whilst statistics of other forms of agricultural produce show a decrease, those of stock show an increase—both in number and value. The best breeds of British stock are the best breeds of the world, and in the present volume each variety of stock is treated of by a specialist. Horses, Cattle, Sheep, and Pigs come in for review; and, in consequence, this little work may prove to have the value of giving a comprehensive review of the more characteristic breeds at a given period—at a time when the stock in question has been brought to a high state of perfection. The best points of the best animals are indicated; and the descriptions and advice may be the more relied on as they are the work of practical authorities. Of course, no particular kind of stock can be absolutely "the best breed," but it may be the best suited to a particular district and for particular climatic conditions—and it is in this sense

PREFATORY NOTE

that the phrase is used. The various chapters are credited to their respective authors; and the Editor confidently trusts that the advice given may be found both interesting and helpful, not only to farmers in Great Britain, but to their friends and connections who have carried their farming with them to the various parts of that Greater Britain which lies beyond the seas.

CONTENTS

	PAGE
SHORTHORNS AS BEEF PRODUCERS	1
SHORTHORNS AS MILKERS	9
HEREFORDS	19
THE SCOTTISH BREEDS OF CATTLE	30
CHANNEL ISLANDS CATTLE	43
THE KERRY AND DEXTER-KERRY BREEDS OF CATTLE	52
DEVON AND SUSSEX CATTLE	61
HEAVY HORSES: BREEDING AND MANAGEMENT	73
SHIRE HORSES	84
LIGHT HORSES: BREEDING AND MANAGEMENT	94
SHEEP	103
PIGS	119

THE BEST BREEDS OF BRITISH STOCK

SHORTHORNS AS BEEF PRODUCERS

By Professor Sheldon

THE development and distribution of the shorthorn breed in modern times form the most remarkable phenomena in the career and history of domesticated cattle. Little more than a century ago the breed was local: it is now cosmopolitan. Before the middle of this century it had become the prevailing breed in the northern and midland counties of England, and was rapidly establishing itself throughout the Eastern States of the great Republic of the West. Other breeds melted away before, or became absorbed in it, leaving hardly a trace behind. The prepotency of shorthorn sires finds no parallel in other breeds, so far at all events as extent is concerned. To have obliterated the longhorns in England, save where special efforts have been made to preserve them, is an achievement such as no other breed than the shorthorn has accomplished. The longhorns were famous a century ago, not in England only, but also in Ireland, where they have likewise gone down under the shorthorns. Similarly, the nondescript cattle of America are giving way before them, even as the Indians have given way before the white men. Grafted on the descendants of the Andalusian cattle, which three centuries ago the Spaniards introduced, the shorthorn characteristics quickly became predominant, and the Spanish are disappearing.

Readers are no doubt familiar with descriptions of huge steers and heifers that were the wonder and admiration of our ancestors a century ago. Indeed, it is not uncommon even nowadays to meet with engravings and mezzotints of those extraordinary cattle, in various old houses in country districts. Perhaps the most famous of these was the "Durham ox," an animal bred by Charles Colling, and sired by the famous bull "Favourite." At five years old this ox weighed 27 cwts., and his carcass was computed to weigh 168 stones of 14 lb. This animal was exhibited in many parts of the country during a period of six years, and this in the pre-railway days. No less a sum than £2000 was offered for him, for exhibition purposes; and it may be assumed that Mr. Day, his owner, made a good thing out of exhibiting him all over the country. At eleven years old his hip-bone was dislocated, and after a time he was killed, his carcass weighing 165 st. 12 lb., in addition to which there were 11 st. 2 lb. of "tallow," whilst the weight of his hide was 10 st. 2 lb. At the same time, Robert Colling bred a white heifer, which became second only in fame to the Durham ox bred by Charles. She, too, was exhibited as a phenomenon, and at four years old her carcass weight was estimated at 130 st. There are records to be found in books of the period of various other splendidly developed specimens of the breed, and to these female animals is no doubt owing, in no small degree, the rapid distribution of the breed in districts and counties beyond the valley of the Tees and the county of Durham.

The brothers Colling were credited at the time with being foremost improvers of the shorthorn; but Allen, the American historian of the breed, seems to take exception to this. Contemporaneous records, however, may be taken to embody the opinion of the period; and, in any case, the conflict of modern opinion with that of the period cannot, in a question like this, lead to practical results of much value. "Whatever had been the merits of the Teeswater cattle, it is certain that Mr. Colling improved them," says Youatt. Nor was this improvement effected by chance, as was asserted; for Colling went to Dishley to learn what he could from that prince of breeders, Robert Bakewell, who had made astonishing improvements in longhorn cattle and Leicester sheep. It may be expected that Colling went a good deal on Bakewell's lines, namely, selecting and mating together animals possessing the qualities he wished to develop and maintain. "He found the Teeswater, like all other extravagantly large cattle, frequently of loose make and disproportion," says Youatt, "and his success resulted from a deliberate and well-considered plan." It appears that a part of this plan was to reduce the size in order to improve the symmetry of his cattle; and yet, for all that, he bred the "Durham ox." No doubt there was an element of chance in Colling's success; but then he was a man who would seize a chance and make the most of it. His chance was to find the bull "Hubback," to whom may be ascribed the extraordinary improvements which were quickly accomplished: this was the "chance" which Colling saw and seized, and what he made of it everyone knows.

The dam of "Hubback" became so fat when put on good land that she did not breed again, and her son had the same predisposition towards obesity to such a degree that he was useful as a bull only for a short period. The quality of his flesh, hair, and skin are said to have been seldom equalled, and it is from him that a large proportion of pedigree shorthorns have derived, in a primary sense, their tendency to lay on flesh rapidly. This bull, indeed, was the pebble at the spring which turned the current of the stream. And yet he himself sprang from the gutter, so to speak; for his dam belonged to "a person in indigent circumstances," who "grazed his cow in the highways." A child of the lanes, like the gipsies, he had a sound and vigorous constitution, and an uncommon aptitude to fatten when good food and plenty of it was within his reach. Whether or no the pedigree shorthorns would have come to possess such a marked aptitude to fatten if the influence of "Hubback" had not been introduced into the breed at the outset, we cannot now, nor could we ever, determine; but it is at the same time true that "Hubback" and his dam had such a powerful tendency in the direction of obesity that they prematurely ceased to procreate.

This is even now a fault among their descendants,

many of which—females particularly—either do not begin to breed, or cease at an early age. It is said that a man in America made money by purchasing pedigree cows that would not breed, or had left off breeding, and putting them to work at the plough. Discipline of this sort restored the procreative functions to action, and the animals again became valuable.

One of the most successful farmers I have ever known used to follow a plan which is quite unusual, though not perhaps unique. Though milking some seven or eight and twenty cows, he did not aim to breed any for his own dairy herd. Living in one of the most favourable districts in England, he bought each spring as many promising milkers as he wanted,— heifers, as a rule,—and all his cows were served by a pedigree bull chosen from a herd noted for soundness of constitution and amplitude of flesh. The whole of his calves, male and female, that were considered suitable were kept thriving as rapidly as might be advisable, never losing their "calf flesh," until at two or two and a half years old they were fit for the butcher, and worth £20 apiece or more. There is in this practice an element of novelty which, however attractive it may be, is not likely to become general among dairy farmers, for it means unstinted outlay, and intensive farming to an extent that would alarm most tenant farmers nowadays; but at the same time, such an instance, prolonged over many years, serves to illustrate the confidence which may be placed in the early maturity and aptitude to fatten of the shorthorns, and in their suitability on the one hand for dairy cattle, on the other for breeding calves specially designed for beef at as early an age as may be practicable, while at the same time they lend themselves conveniently to any modification of practice which may be possible or desirable in dairy farming.

It is indeed, at all events in part, their cosmopolitan adaptability which has recommended the shorthorn alike to the dairy farmer and the grazier of the British Islands, and to the "dairymen" and the ranchers of America. Milk, beef, early maturity, quick fattening, adaptiveness to any reasonable soil and climate, prepotency, fecundity, vigour of constitution—these are the merits of the shorthorns which have caused them to follow the Englishman to every country where his tongue predominates, and to some where it does not. Away from the fertile valleys of Nova Scotia to the rolling foothills of the Rocky Mountains, on the vast cactus prairies of Texas and the parched plains of Mexico have I seen them quite at home and flourishing, and in some places revelling in boundless freedom. Yet again have I seen them, big, handsome cattle, pitiably sad to look upon, flood-drowned in the swampy bottoms of Missouri, and floating dead on the great billows of the Atlantic—thrown overboard from some ship in distress.

In the Canadian ranching districts the winters are very severe, as a rule, but the shorthorns winter well out of doors, in a semi-feral state, if only they can get food enough. Losses have at times been heavy among them, but these were the result of scarcity of food rather than of the severity of the winters. So long as the cattle had to depend on their own exertions for food, it is not surprising that many of them should die of starvation when the grass was buried for months under half a yard of frozen snow. But now that hay is put up in ricks, and fed to the cattle in bad weather, the losses are very small. It may be said, indeed, that shorthorns will thrive throughout the north-west of Canada if only food for winter be provided; and, so far as this aspect of the case is concerned, we may well ask what would become of them in many parts of England but for the provision of winter forage. I have seen large herds of cattle in Alberta, which would be a credit to any stock market in England, fed solely on the apparently coarse and innutritious herbage of the prairies. And yet this ranching country is some 3500 feet above sea-level, and in a latitude which, on the Atlantic coast, admits of nothing but barrenness and wild desolation. This latitude, however, is that of the British Islands, though the climate is not so temperate as here. But in any case, dissimilar as the climate and herbage of the two countries are, the shorthorns, to all appearance, lay on flesh and fat equally well in both.

The only other breeds that can even remotely be looked upon as possible rivals of shorthorns in the great ranching countries are Herefords and polled Aberdeens, but these never can be their rivals in dairying districts anywhere. It is, no doubt, true that the Herefords yield beef the quality of which is by some people considered a little superior to that of the shorthorns, and the same claim is sometimes made for that of the polled Aberdeens. It may or may not be true that the shorthorn beef is a trifle inferior; this to a great extent is a question of taste. But in any case the shorthorns will, as a rule, and under equal advantages, make more beef at an early age and in a given time than either of the others. It has been urged that the polled Aberdeens are hardier than the shorthorns, and this no doubt is true; but as the shorthorns flourish in climates as disparate as those suggested by the countries I have named, it would appear that they are hardy enough for all practical purposes, and for any reasonable soil and climate.

Many discussions have taken place on the question of breeding for beef and milk combined, and some men declare that it cannot be successfully done, whilst others aver that it can. In his account of the shorthorn breed, before quoted, Youatt says: "It is evident that bad milking in a breed of animals which were ever distinguished as good milkers is not a necessary consequence of improvement in the animal in other respects, but a consequence of the manner in which such improvement is pursued"; and he asserts that "improved shorthorns, inferior to none for the grazier, may be selected and bred with the most valuable dairy properties." It is at the same time true, in all probability, that if milk may be sacrificed to beef in one case, and beef to milk in another, the results of breeding for one special object instead of two may be higher than where the two are sought combined. But why should either be sacrificed? The wisest plan is surely that which aims at both; for both are necessary to the best possible breed of cattle, and the best possible is what we want in these days of sharp competition and diminishing profits.

SHORTHORNS AS MILKERS

By Professor Sheldon

The subject of lactation in dairy cattle is one on which a great deal has been written and spoken, and some of the men who have made a study of it do not seem disposed to admit shorthorns into the front rank of milk-producing cows. A few, indeed, have gone so far astray as to aver that they are essentially beef cattle, and not milkers. These persons are, however, chiefly, if not exclusively, Americans with whom the Jerseys, the Ayrshires, and the Holsteins are the prime favourites at the milk-pail. Be this as it may, we in England are well enough content to allow others to think and say what they like so long as we are permitted to believe that, take them all in all, the shorthorns may with confidence be placed as one of the best half-dozen milking breeds to be found anywhere, and that as all-round cattle they cannot be surpassed, if indeed equalled, by any other breed in the world. We may venture to suspect that American opinion on this point would have been more favourable if the shorthorns had been indigenous to the great continent instead of the little island. But the pride of the bovine kingdom is not indigenous to the New World any more than is the white man himself; and it is not yet demonstrated that the soil and climate of the New World will permanently maintain the characteristic vigour and quality of the shorthorn, whatever they may do for the white man, unless fresh blood be constantly imported from the Old Country.

If we draw attention to the remarkable facts that the shorthorns have, during the present century, become more widely and numerously diffused than any other breed throughout the dairying districts of England and Ireland, not to mention Scotland and Wales; that over large areas of country they have displaced or are displacing other breeds, against which process there is at present no reaction worth the name—enough is said to show that the premier breed possesses superior and commanding merits as milk producers. All this is in no respect the result of mere undesigned and accidental fashion, but to the well-ascertained fact that as milkers, as breeders, and as fatteners shorthorns are superior to the breeds which they have displaced. Less than a century ago, the quaint and interesting longhorns were prevalent in most of the midland counties of England, and the shorthorns were confined almost entirely to their native habitat in Yorkshire, Durham, and Northumberland. But for many years past the longhorns have existed only here and there, in isolated cases, more as curiosities than as practical dairy stock; that is to say, a few noblemen and gentlemen have clung to the breed despite the prevailing tendency of the period. This vast transposition of breeds supplies us, in the domain where the influence of man has been exercised, with a marked instance of the "survival of the fittest,"—the most suitable and, therefore, the most profitable for dairy farming purposes.

The Teeswater cows, from which sprang the celebrated shorthorns of to-day, were noted centuries ago as deeper milkers than any other breed in the British Islands—so, at all events, ran the old-world tradition concerning them. An ancient record, said to be still preserved in Durham, states that cattle of great excellence existed in that country in the fifteenth century. Upwards of two centuries ago, according to the late L. F. Allen's valuable history of the breed, "there existed fine stocks of shorthorn cattle in Durham and Yorkshire, which are described as 'large milkers.'" Great attention was paid to these cattle in that early period by such owners as the Aislabies, of Studley Park, Yorkshire, and the Blacketts, of Newby Hall, Northumberland. In those early times, therefore, the shorthorns were highly esteemed, not only as growing into fine carcasses of beef, but as producers of large quantities of milk. The brothers Charles and Robert Colling, who did so much to establish the fame of shorthorns, and whose names are inseparably connected with the early fortunes of the breed, were careful to cultivate all the best qualities of which the breed is capable, and among them, we may assume, the property of copious milking. There are many other famous breeders, contemporaries of the Collings, whose cattle partook of the all-round improvements that were accomplished by the careful selection and mating of animals in which existed the qualities it was desirable to develop and perpetuate.

To these early breeders and improvers of the

shorthorns much credit is due. In a period when no great inducements to excel existed, they created the inducements by improving the cattle. They laid down the foundations upon which all subsequent shorthorn achievements rest. They evolved a superior class of cattle, which in time commanded a superior market. No doubt it is true that the raw material they worked with was potentially excellent, but this excellence required skill and judgment to develop it. These old shorthorns were known to be excellent milkers—the fourteenth century sculpture of a shorthorn cow and of milk-maids with milk-pails, still existing in a niche on a tower of Durham Cathedral, denote this in symbols of stone. But they were coarse, big, ungainly, and wanting in compactness and symmetry, according to the ancient traditions. Milk, however, they had in abundance; and the desired improvements were physical as well as functional, aiming to secure evenness of form, early maturity, aptitude to fatten, quality of flesh, and beauty all round. These were the lines on which the early process of improvement was conducted, and in course of time the shorthorns attained a degree of fame surpassing that of all other breeds, save Robert Bakewell's longhorns. The greatest of breeders was Bakewell, and we may regret that his skill was not employed on the shorthorns, for they would have won him greater credit still, in milk as well as beef.

The shorthorn improvement moved steadily onward in all the succeeding years. The fame of the Collings and others was emulated by Bates and Booth and a hundred more. Unfortunately, however, many of the subsequent breeders of pedigree shorthorns began to sacrifice milk to form and beauty, and in this way some of the famous families and tribes lost their reputation for the dairy, and injured that of the breed in general. That this was a disastrous policy modern events have abundantly proved. Not all of the great breeders, however, followed this line of breeding, and the cows of Bates, Whittaker, Knightly and many other men were not suffered to lose their birthright of milkiness. During the period of diseased inflation of shorthorn values, when 1000 guineas were not looked upon as a very remarkable price for a heifer or bull, milk was regarded as having no practical value, and as being a hindrance to the development of other qualities. The cows were expected to breed, but not to provide milk enough for their calves. Commonly, however, the system of close in-breeding, the constant pampering with food, the unnatural treatment to which some of these blue-blooded cattle were subjected, made them barren and unprofitable—calfless, milkless, a travesty on their breed and sex. As a matter of fact, the calves bred by these aristocratic mothers required wet-nurses to rear them, which may be regarded as a negation of maternity.

The chief end and aim of a cow's existence is, or ought to be, the production of a calf, and of some two and a half to three tons of milk per annum, say for four or five years, and then to turn herself as expeditiously as circumstances admit of into a carcass of good beef weighing about one-third of a ton. This is what a practical shorthorn cow tries to do, if she gets anything like a reasonable opportunity; but it is what many pedigree shorthorns have not been allowed to do, save and except the last part of it. It is not these animals' fault, but their misfortune, that their destiny has been frustrated in this manner. It was inevitable that the time should come when customers among practical dairy farmers must be found for many of the young bulls of the pedigree herds. The supply of these fashionable young sires soon became greater than the demand for them within the aristocratic circle, and it was desirable that the surplus should mate themselves in middle-class families, so to speak, in the shorthorn world as an alternative to emigration and castration. Practical dairy farmers, however, require bulls from milky mothers, or else the connection ends in disaster. And again, they want bulls that have not been pampered from their birth and forced into unnatural maturity. Enormous mischief has been done by neglecting these considerations. One excellent breed of unpedigreed shorthorns I have known to be ruined in this way, by the introduction of a high-priced bull, a bull that had been forced and pampered with food into premature flesh and sleekness, which could not be maintained when he came to be lord of the harem. These bulls, like other things, are too much "got up for sale."

As I have said before, the question of yielding a copious supply of milk has been too long neglected, purposely neglected, by some of the breeders of pedigree shorthorns. In the halcyon days of three and four figures these men could well enough afford to treat the lacteal function in a cow as a merely subsidiary and subordinate characteristic, more honoured in the breach than in the observance. But since the figures have come nearly down to the intrinsic level, with a continued tendency to decline still further, breeders' eyes have been opened to the true value of a reputation for milk. The herds in which this reputation has been wisely and constantly maintained have found customers freely enough for surplus stock in recent times, whilst others have gone a-begging. This shows a healthy return to first principles, and we may safely conclude that the lesson will not be forgotten by those to whom it has been sharpest. The future, therefore, promises to add more than the modern past has done to the ancient fame of the shorthorns for milk; the pedigree shorthorns, of course, being here alluded to, for it is they (or some of them) which alone were permitted to lose it during the period of inflated values. It may well be doubted whether the highly fashionable period through which the shorthorns have been passing has been a really good thing in the interests of the breed, and whether the gambling spirit in which some men took them up has not done mischief which will take long years to efface.

When we regard the breed outside the strictly pedigree part of it, and come to herds that have been bred for milk,—that is, for practical dairy purposes,—whether by an occasional or even a regular infusion of pedigree blood, or even without, we find that shorthorns can render as satisfactory an account of themselves as any other breed in reference to milk. Probably, however, such herds are not to be found here, there, and everywhere, and it is certain that they are not, and have not been, nearly as numerous as they ought to and might have been. The late Mr. E. C. Tisdall, of Holland Park Farm, a much respected London dairyman, had an excellent herd of shorthorns, fine handsome cattle and grand milkers. Mr. Tisdall was from the first one of the best friends of the British Dairy Farmers' Association, and to its *Journal*, in 1878, he contributed an article on the "Improvement of Dairy Cattle," which contains many sound and practical remarks on the subject. In this he gives remarkable figures relating to the milk-yield of various cows and heifers, some of them his own. He is very decided as to the disadvantages of feeding and training for shows; one heifer of his, so trained, won the prize at the Chippendale Show, "but in winning that she lost her family's reputation (for milk) and her own usefulness." He used good bulls from well-known milky herds, some of them pure bred; and in 1877 he had ten heifers, by a pure-bred bull, each of which on an average yielded 868 gallons of milk in twelve months. Perhaps the most remarkable instance on record is that of "Old Strawberry," a cow belonging to Mr. Lakin, of Beauchamp Court. This extraordinary milker averaged 1050 gallons per annum, or per ten months, during no less a period than *fifteen years.*

It is well known that in Cumberland and Westmorland may be found in many places as good a class of well-bred but unpedigreed dairy shorthorns as in any other county. In Yorkshire, too, markedly in the Holderness district, and also in Lancashire in the famous Fylde district, in Derbyshire on the limestone soils, and in Cheshire on the Keuper marl, as well as in various other counties, and on soils disparate in character, the rank and file of the breed are seen in full possession of the qualities for which the Shorthorns are famous—milk, mellowness of skin and flesh, amplitude and symmetry of form, early maturity, hardiness of constitution, fineness of bone, fecundity, and so on. These are the sort of cattle best suited to the needs of dairy farmers, subject to the selection of the fittest. Any new beginner in dairy farming, if he be a fair judge of cattle, may soon get together a herd good enough to start with; if he have doubts of the maturity of his own judgment, he will do well to give a commission to a trustworthy dealer. Inferior milkers should be weeded out of the herd as soon as possible, and all heifer calves from good milking dams should be kept for use in the future. It is of the greatest value to keep a record of the yield of milk by every cow, and the simplest way of ascertaining the quantity is to have milk-pails which are graduated in pints. The keeping of such a record of both quantity and quality is surprisingly interesting as well as instructive, and it is the only reliable way of deciding which of the cows are worth keeping and which ought to be run barren and be fattened for the butcher.

In all districts where the land is fairly good, shorthorn cattle will be found as good as, if not better than, any other sort, for breeding, for milk, and for sale either as barreners or autumn calvers. But it is always advisable to keep up the quality of the herd by using well-bred bulls, be they pedigree or no. Milk is the chief consideration—plenty of it, and of good quality. The milk yielded by Shorthorns is not equal in quality to that of several other breeds—to wit, Jerseys, Guernseys, Kerries, and Devons; but it is good enough to pass the standard of the milk trade, namely, 11 per cent. of solids, and to leave a margin to the good. It is only too common for dairy farmers to breed from animals that are inferior, or at the most only middling, for milk; and whilst this is a practice among a large proportion of farmers, the Shorthorns as a breed cannot attain the high average standard of which they are potentially capable. If, on the contrary, it were practicable to induce all farmers to breed from animals in which copious milking is hereditary, we should soon perceive a marked improvement among Shorthorns, which are so ready to respond to it; also amongst other breeds, though probably not in so great a degree. Shorthorns no doubt offer scope enough for selection, for there are good, bad, and indifferent among them; but they offer also a field for improvement, more promising perhaps than we shall find elsewhere.

HEREFORDS

By Alexander Macdonald

A snack of historical reading on a pet theme is enjoyed by most people. It is all the sweeter, perhaps, when freely intermingled with information of a less permanent nature, designed to impart freshness to the mind and relief to the eye. In this practical age, indeed, it may be doubted whether the tincture of freshness, by which the present position of the subject under review is more clearly reflected, is not better relished and appreciated than the more enduring materials interwoven with it. On this assumption, at any rate, and in obedience to a strict personal preference of up-to-dateness in all such matters, the cart of custom shall precede the horse of habit in the observations bespoken from me upon Hereford cattle.

At the Birmingham Fat Stock Show of 1894 the breed cup was won by a Hereford steer under three years of age, scaling $18\frac{3}{4}$ cwts.; whilst a Hereford-Aberdeen-Angus heifer carried off the special prize for the best cross-bred animal in the hall. At the Norwich Exhibition, too,—a national meeting, like the former,—the Herefords won champion honours against all breeds, vanquishing the Red Polls on their native heath, and practically carrying all the prizes available for them. Nor is this array of achievements phenomenal. The white-faces may be said to have championed every fat-stock show of national importance in the world—English, Australian, American, and French. The coveted trophies of the Islington and Birmingham Exhibitions have repeatedly fallen to it; and at least one Elkington Cup—which has to be won thrice by the same breeder to become his own—has been won outright by a Hereford. The first Elkington Cup ever finally wrested from the vortex of covetous competition in Bingley Hall was won by Mr. John Price, one of the most successful breeders of Herefords in its native county; and the very fact that the Hereford was the first breed to capture the valued trophy is significantly creditable in itself.

Beef-making is the pre-eminent function of the Hereford. Nevertheless it has many virtues to explain its surpassing popularity as a farmer's and grazier's breed. Although no official record of genealogical descent was kept prior to 1845, when the Herd Book was introduced, the breed appears to have been carefully bred as a distinct race of stock before the century began. Its colour then was not so uniform as it has latterly become; neither, of course, was such precision observed in perpetuating certain strains; but it appears to have shown such remarkable hardiness of constitution and fecundity as to suggest the propriety and importance of keeping the breed as distinctive as possible. The obtrusive white face and chest and red body, now universally characteristic of the pure-bred Hereford, dates, according to the author of *Saddle and Surloin*, from the birth of a bull calf with this peculiar marking, at the end of the last century. Breeders fancied the new coat, and endeavoured to emulate it, with the result that the new colour of the Hereford has long since become as fixed and uniform as that of the magpie. The breed is pronouncedly pure—unite it with almost any other breed of cattle you may, and the produce will be almost predominantly Hereford. This is not always so, however, when mated with pure-bred Shorthorn or Aberdeen-Angus cattle of strong constitution; but no other breed, even under equal conditions, will leave more emphatically its impress upon its progeny, whether they be pure or hybrid. The white face is inextinguishable; while it transmits concomitant, if less conspicuous, features with almost equal certainty. Take, for example, its remarkable massiveness—length, depth, and breadth of carcass, expansive chest, and deep protrusive rounds. These distinctive points invariably accompany the diffusion and transmission of Hereford blood, in degrees proportionate to its prepotency in the union.

A white face on a cross-bred animal will atone for many faults. Although the carcass of the Hereford yields less internal fat than those of some other breeds, or than the butcher appreciates, it is rich in the higher qualities of beef. I question, indeed, if any other breed, not excluding the popular Shorthorn, or the solid, fine-boned Angus, carries its meat so largely on its back as does the Hereford. Everything it eats goes to the manufacture of the profit for the producer, and it has been clearly demonstrated of recent years, that it does not waste time in the process of fattening. The weigh-bridge of the leading fat-stock shows in

England and in Chicago have established beyond all question that the Hereford is by no means the slow maturing animal which it has been by adverse critics characterised. Young steers of pure Hereford blood have appeared at the Christmas shows on both sides the Atlantic of late years with a very creditable daily gain record in weight of $2\frac{3}{4}$ lb. since birth. Occasionally the Shorthorn has made a similar record, but, taking the young classes in these fat-stock shows as a whole, Herefords distinctly top the list as regards weight for age. The class of animal which has run this record closest, if it has not in the general average eclipsed it, is the Hereford cross-breds—Hereford-Shorthorn, or three-part Hereford bullocks. This class of stock has become very popular in many parts of the world, especially in the United States of America and in Australia. American breeders have a pronounced preference for what they call "Hereford grade" cattle; and it is no uncommon occurrence to find leading breeders in the West challenging the supporters of any other breed to produce a given number of cross-bred cattle, of the highest and most advantageous grade procurable, equal to a similar group produced by even Texan or native cows of the West, and sired by a Hereford bull.

As a ranch breed, the Hereford is probably unequalled. Its hardiness, combined with its singular aptitude to fatten upon natural foods, have begotten it a position of supremacy in the wild ranching West. The British breeds of cattle have their respective advantages at home, adapting them pre-eminently to certain districts of the country, but when they meet on the Western prairie they meet on common ground, and their respective merits are more evenly put to the test. The prize of this contest, however, there can be little doubt, has fallen to the white-faces, for animals of Hereford characteristics are most eagerly sought after in the leading beef markets of America. Nothing arrested my attention more sharply, while on a tour through the States, than the great preponderance, at all the leading American shows and fairs, of high-grade Herefords—Hereford crosses of select breeding. Having accepted repeated invitations to aid in the distribution of the prizes at several of these meetings, including the Chicago Fat Stock Show, I had ample opportunities of examining the animals closely. Such a magnificent string of cross-bred heifers under two years old as came before me at Chicago I have never seen, of any breed in any country. Well-grown, heavily-fleshed, and remarkably well-matured, they gave excellent testimony to the triumphant success of the Hereford breed as the medium of first-class rent-paying material, capable of being matured early if desired. And, furthermore, the animals, both pure and cross of this breed, which received the principal prizes, on foot, I had the advantage of examining in carcass. Americans are ahead of us in this important arrangement, in that they wind up the Chicago meeting every year with a block test by which judges, breeders, and visitors are enabled to see both the exterior and interior of the most successful animals, the process of killing and dressing only occupying a few minutes—$12\frac{1}{2}$ minutes in the case of the most expert butcher. The Hereford carcasses dressed beautifully, and compared advantageously in sectional views with those of other breeds represented on that occasion, excepting the carcass of a prime little Sussex steer that received the first prize on foot, whose meat was so delicately-marbled and fine in colour and quality as to secure the first prize on the block.

Whether we take the Hereford breed in its purity or in union with other strains of stock, there can be no doubt of its supremacy as a grazier's breed. It will fatten to perfection on grass, which it exists upon to a much larger extent, perhaps, than any other beef breed. Long has it been recognised as producing the best grass-fed beef sent into London market, while it also responds effectively to less natural diets. I refer to the severe dietetic strain to which prize-winners at the English shows, both of breeding and fat stock, are subject in the process of preparation for the show-ring. They stand the pressure well, and never fail to accumulate flesh in the right parts. All these virtues are natural bequests of a carefully-bred parent stock, which, while it has proved itself capable of producing animals that mature early under special treatment, has long had the reputation of being a breed that improves and develops for an unusually large proportion of its life; that, living under the most natural conditions, is prolific; and that continues profitable to a long age. The cows live on the pasture for the greater part of the year, and the young stock, not intended for show purposes, in many cases live out of doors all the winter. They are essentially grazing cattle, and are better adapted for a large extent of country, outside the county from which they take their name, than any other breed. This accounts for the multiplicity of the pure-bred herds of Hereford cattle found in Shropshire and other neighbouring counties; and its popularity and success as thrifty and rent-paying cattle is forcibly witnessed to by the dimensions to which the Herd Book of the breed has lately been developed. The volume for 1894 contains the names of upwards of 380 members, and bears evidence in every respect of increasing vitality and progress in the interests of the white-faces.

Hereford breeders are sometimes twitted with the remark that their favourite cattle are bad milkers. There is abundant reason for knowing that they are immensely superior on the block than at the pail; but this is no reflection upon the natural qualities of the breed. Little or no effort has been made—at any rate, not that effort which other breeders have made in the interests of their stock—to develop the milking properties of the Herefords. They have been bred and reared almost exclusively for beef, yet they generally nourish their calves well, with, in many cases, something to spare. In fact, there were some years ago, if there are not at this day, herds of Hereford cattle principally devoted to dairying. In a lecture delivered before the Brecon Chamber of Agriculture, Mr. Duckham stated that a Hereford herd had been established for thirty years in Dorsetshire, specially for dairy purposes, adding that the breed was rising in favour in that county by virtue of its excellent dairy properties. The owner of the herd specially mentioned by Mr. Duckham wrote: "In proof that they are good for milk, I let out a hundred cows to dairy people, and if I buy one of any other breed to fill up the deficiency, the dairymen always grumble, and would rather have a Hereford of my own breeding." Mr. Oliver, of Cornwall, also a breeder, stated to Mr. Duckham that, in his experience, if Herefords were bad milkers, this arose, not from any constitutional defect, but rather from mismanagement, or neglect to cultivate the milking properties of the breed; adding that "my cow 'Patience,' bred by Mr. J. G. Cook, Moreton House, Hereford, has yielded 14 lb. of butter per week; and 'Blossom,' bred by Mr. Longmore, Buckton, Salop, gave 22 quarts of milk, yielding $2\frac{1}{2}$ lb. of butter per day."

Milking and breeding cows, when taken to the steading for the winter, receive little else in the shape of food than turnips and straw. The principal variation of this rule is when hay is substituted for the straw, as sometimes happens if the animals seem to require a little better food, or when hay is plentiful and cheap. Calves, as a rule, arrive from the 1st of February till the end of April, and run with the mother in the fields throughout the summer. If intended for sale, the youngsters get a little cake in addition to the commoner ingredients, to assist fattening during their first winter; and they are usually ready for the market by the end of the following summer, when they make prices ranging from £12 to £20. Many breeders bestow a second winter's feeding, which is somewhat richer than that of the first, upon the steers, selling them as prime beef in the spring.

But it is not my purpose to dwell at any length

upon the practical management of the Herefords. Reference may be more fittingly made to the successes of the Hereford in the showyard and saleyard during the past few years. The breed has long had the patronage of influential and energetic agriculturists, including royalty and country gentlemen, and has therefore been well represented in the show-ring. In other words, it is in the hands of people who can afford to develop its merits to the greatest possible advantage. And in this connection anyone who is at all acquainted with its history will think of Her Majesty the Queen, Lord Coventry, Mr. Rankin, M.P., Mr. Arkwright, Mr. Tudge, Mr. Taylor, Mr. Cook, Mr. A. P. Turner, Mr. Evans, Mr. Hughes, Mr. Fenn, Mr. Godsell, Mr. Russell, Mr. Green, and Mr. R. Palmer, as prominent winners at shows of recent years. Her Majesty's herd at Windsor, so successfully managed by Mr. Tait, is one of the best in the country, and has produced prize-winners in both breeding and fat stock shows for years. Lord Coventry has had a remarkably successful career as an exhibitor of Hereford cattle, and has bred many animals which have added lustre to his fame in this and other countries. Breeding show cattle has attained to a remarkable degree of perfection. This will be freely borne out by all who have been regular visitors to the Royal Agricultural Society's Show, especially those who can recall such magnificent specimens as Lord Coventry's "Good Boy" and "Rosewater," both of which left a record of unbroken showyard success.

Turning to the sale-ring, the most luminous event in the history of the breed recurs to memory. The name of "Lord Wilton," which made the enormous figure of 3800 guineas, from an American breeder, and for which Mr. Rankin, M.P., was the last unsuccessful bidder, stands out in pre-eminent distinction. Such extraordinary figures as this are not to be expected every day; yet some very high prices have been paid for specimens of the Hereford breed in later years. Good cattle of known and cultivated strains command remunerative prices; and in this breed, perhaps more than in others, a great deal of business is done by private treaty. A few herds are wholly or partially dispersed by public auction during the summer months; but the event which practically determines the character of the year in Hereford sales is the great annual fair held at the county town in the autumn, when several thousand cattle of all ages are exposed.

It is, of course, impossible in a single chapter to fully exhaust such a subject as that of Hereford cattle. I append, in conclusion, the following statistics indicating the relative measurement of well-known prize Hereford and Shorthorn cattle exhibited together in an English showyard and measured on the spot :—

Bulls.	Shorthorn.	Hereford.
	ft. in.	*ft. in.*
Girth round shoulder	8 0	9 5
Shoulder to rump	4 10	5 2
Shoulder to hip	2 10	3 0
Loin to flank	2 6	2 6
Across the loin	1 10	2 0
Hip to rump	1 9	2 0
Hip to hock	3 3	3 0
Girth round flank	8 7	8 11

HEIFERS.	Mr. Acker's Shorthorn. 2 yrs. 6 mos.	Herefords.	
		"Kathleen" 1 yr. 9 mos.	"Rosa" 2 yrs. 4 mos.
	ft. in.	*ft. in.*	*ft. in.*
Girth round shoulder	7 0	7 0	7 8
Shoulder to rump	4 7	4 5	4 5
Shoulder to hip	2 9	3 1	3 2
Hip to rump	1 9	1 8	1 9
Across the loin	1 9	1 9	2 1
Hip to hock	2 10½	2 8½	2 10½
Hip to flank	2 0½	2 2	2 1½
Girth round flank	7 0	7 6	7 11

THE SCOTTISH BREEDS OF CATTLE

THE ABERDEEN-ANGUS

By C. Macpherson-Grant

This variety of British live-stock has within the last decade so enlarged the area of its usefulness, that it is scarcely necessary at this date to point out that it is black and hornless. It had its origin in the lower districts of Aberdeenshire, where it was for long known as the Buchan Humlies, and in part of Forfarshire (or Angus), where it was locally called the Angus Doddies, the two words "Humle" and "Dodded" being widely used in Scotland in those days to express the absence of horns. Subsequently the two names became merged into one—"Polled"; but as there exist two other and absolutely distinct varieties of hornless cattle in the British Islands,—the Galloway in Scotland and the Red Polled in Norfolk and Suffolk,—the title of Aberdeen-Angus has now by universal consent been adopted. The points of a good Aberdeen-Angus bull may be shortly given as follows :—Black (a very little white on the underline behind the navel is admissible, but a white cod is most objectionable); forehead broad; muzzle fine; distance from eye to nostril of moderate length; eye full, mild, and expressive; there must be no horns or "scurs," which are short, loose knobs not adherent to the skull. Well-laid shoulders, which on the top should be moderately broad and in line with the back; ribs well sprung from the back, arched and deep. The body should be deep, broad, and symmetrically placed inside the legs, which should be short, straight, and squarely placed, with fine, clean bone; the hooks (or hips) must not be too wide, and there must be no hollow between them and the rump. The tail should be fine, and come neatly out of the body on a line with the back. The flesh should be even without any sign of patchiness, and the skin of moderate thickness (not papery) and abundantly covered with thick and soft hair. *Mutatis mutandis*, this description will do equally well for a cow, but a white udder is not so objectionable. Although black is now the colour recognised, it was not always so, and half a century ago it was not unusual to find good specimens red, yellow, or brindled. It is now very rarely indeed that atavism occurs, and red calves are scarcely ever produced.

The first systematic improver was undoubtedly Mr. Hugh Watson, of Keillor, Forfarshire, and his example was shortly after imitated by Mr. McCombie, of Tilly-four, in Aberdeenshire. About the same time, Mr. Bowie, Mains of Kelly, and the Earl of Southesk (then Sir J. Carnegie), did much for the breed. At the present day breeders and owners of excellent herds are numerous, and it would be an invidious task to select any few gentlemen for mention here. The breed has proved its worth in so many hard-fought contests at Islington, Bingley Hall, and elsewhere, that many

Englishmen and Irishmen have joined the ranks of its admirers, and one or more herds can now be found in many English counties and all over Ireland. So much so that several district societies in England give prizes for Aberdeen-Angus cattle, and recently the Essex Society, we believe, announced their intention of doing so. The Royal Dublin Society also have classes, and, by means of the Queen's Premiums given to bulls, have done much to enlarge the sphere of usefulness of the breed and its popularity.

Our American cousins and our fellow-countrymen in Canada have taken over to the other side many animals, some of great merit, where they have been substantially increased, and have proved themselves on the ranches to be as good "rustlers" as any other variety of British bovine stock. Aberdeen-Angus cattle have also found their way to Australia and New Zealand, to the River Plate, and even to Egypt. Indeed, they give promise of becoming as cosmopolitan as the Shorthorn. One reason, and it appears to us a very cogent one, is the prepotency of a purely-bred Aberdeen-Angus bull. It has been demonstrated that quite 95 per cent. of his calves will be black *and hornless*, no matter what females he may have crossed; and in these days of transit by rail and steamship of such vast numbers of cattle, the latter merit is an obvious one, for it does away, by nature's method, with the need of the more or less painful process of dishorning, which in this country has given rise to much feeling and litigation. Anyone who reads the reports of Smithfield markets will usually find that prime Scots are quoted at a trifle more money than other descriptions; and it is well known that the Scots are generally in black skins, although not necessarily purely bred.

It was for some time denied that Aberdeen-Angus cattle could mature as rapidly as Shorthorns, but the idea has now been effectually dispelled; and at Smithfield Show they have frequently proved, both in gain per day and in percentage of carcass, their equality with that of any other breed. Considering, as we ought to do, that in number they fall far short of the most widely-diffused breed of modern days, this speaks volumes for their excellence. We have no hesitation in asserting that the best beef animal of the present day will be found in a cross between the Shorthorn and Aberdeen-Angus, and in practice it will be found that the bull may be of either breed with equal advantage; but if the male be a Shorthorn and his cows Aberdeen-Angus, it is probable that not quite so large a proportion of the calves will be black and without horns; and if a white sire is used, probably many of the calves will be of the blue-grey colour so much appreciated by English feeders. An association has been formed in the United States, and a Herd Book published, for the registration of a "polled" sub-variety of the Shorthorn, and we understand that the gentleman who established this fancy did so by the careful use of Aberdeen-Angus bulls.

The breed have been found quite hardy, even in the severe climate of Wyoming and the North-West Provinces of Canada, and are equally good doers in our own islands. Indeed, there is frequently difficulty in keeping the two-year-old heifers (at which age, under the best management, they are served) from attaining a too great degree of fatness. The ages of Aberdeen-Angus cattle are reckoned from 1st December, and most calves are dropped in that month, January, February, March, and April. Spring sales of young bulls are annually held in many centres, the most important of which are Perth, Aberdeen, and Inverness, and it is found that good, strong bulls are in great demand for crossing purposes, while the best find their way into pure-bred herds. There would appear to be a growing tendency to ignore pedigree, but this, in our opinion, is a great mistake. Breed, tribe (or family) within the breed, and parentage may be considered, in the order named, to control the offspring, and it by no means follows that by mating two first-rate animals you attain nearest to perfection. Heredity may, very probably will, upset your calculation, and this has been found true in the history of all pure-bred animals; to this rule the thorough-bred horse, the longest recorded of live stock, is no exception.

We have made passing allusion to victories of Aberdeen-Angus cattle at Birmingham and Smithfield. Perhaps their most notable success was gained in 1881, when Mr. Walker (manager to Sir W. Gordon-Cumming) sent a pair, little more than two and a half years old, from Altyre to Smithfield. He carried the special prizes as the best male and female; while for the championship of the show the contest lay ultimately solely between these two animals, the fiat of the judges going in the end in favour of the heifer. Since that date we have had the victories of "Luxury" and "Young Bellona," and other successful contests.

At the International Exhibition at Paris, in 1878, two champion prizes of £100 each were offered for the best group in the division foreign to France, and for the best group in the entire show. Mr. McCombie, of Tillyfour, had the distinguished honour of winning both with a beautiful lot of Aberdeen-Angus cattle, and defeating, among others, Lady Pigot's Shorthorns. In the second contest the "runners-up" were a group of French Shorthorns. In both cases it is known that the blacks won by votes of large majorities of the jurors. Such a triumph as this, followed up, as it constantly is, by further successes at our leading shows, has induced many agriculturists to try the blackskins, and we believe they have done so with satisfaction and profit to themselves.

At three of the leading sales of 1895 the averages were as follows :—For fifty-two animals, £54, 13s. 3d.; for thirty animals, £39, 11s.; and for seventy-nine animals, £50, 2s. 8d. At the Perth sale a recent spring one-yearling bull brought 190 gs., and another 175 gs. We mention these facts to show the approximate values of the best animals in recent years, but undoubtedly excellent herds can be founded on much more easy terms.

THE GALLOWAY AND THE AYRSHIRE

BY JAMES LONG

THE GALLOWAY

The Galloway breed of cattle, although black in coat and polled, and thus resembling in the most important outside features the breed just described, is certainly not its equal, regarding it from the point of view of precocity, quality, or form. As in the Angus, white frequently appears on the underline; but there is one distinctive difference between the two breeds, namely, that the black coat frequently carries a rusty or brown tinge. Many Galloways, too, unlike the Angus, present a curly or wavy appearance in the coat, which certainly adds to their character; and there is also a distinct difference in the form of the head, which is emphasised in winter when the coat is fully grown and covers the forehead in curly masses. The name of the breed is derived from the districts of Kirkcudbright and Wigtownshire, which are generally known as Galloway. According to Professor Wallace, who has made a study of Scotch breeds in particular, it is descended from the same stock as the still shaggier, but horned, West Highland cattle, Galloways being described as Highlanders without horns. It is said, too, that when a Galloway is crossed with a West Highlander, judges of Galloway cattle have sometimes been unable to distinguish the cross-bred progeny from pure specimens of the breed. Although a good beef-producer and a hardy mountain breed, the Galloway is not a dairy cow; indeed, it is doubtful whether at all times it provides a sufficient quantity of milk for its own calves, which usually run at its side. In this respect it is inferior to the average Aberdeen-Angus, especially as some cows of this variety have been very excellent milkers indeed. The Rev. John Gillespie, the editor of the *Galloway Herd Book*, has remarked that when cattle of a horned breed are crossed with the Galloway, the half-breeds, in ninety-nine cases out of every hundred, are polled, and he therefore believes that there is no breed of polled cattle so impressive for use in the abolition of the horn. There is perhaps no breed, unless it be the West Highlander, which is so well fitted for exposure to extreme heat and cold; and it is probably the tendency to leave Galloway cattle out at all seasons that accounts for their slower development and their later maturity. Although the hide of the Galloway is not so thin or so soft as that of the Angus and of some more southern breeds, its meat is of beautiful quality and well-marbled, strong store cattle feeding well when brought south and tied up for the purpose. With regard to the economical value of the Galloway, a case may be mentioned in which half a dozen steers were sold to a butcher at 9d. per lb., carcass weight, which was assumed to be 53 per cent. of their live-weight, upon the basis of a weighing immediately before feeding. Three of the animals were Galloways, and three were Shorthorn-Ayrshire crosses of the same age, all fed in the same manner. The Galloways weighed 72, 71, and 75 st. respectively, the cross-breds 74, 59, and 60 st. The estimated carcass weight of the whole was 3061 lb., but the actual weight turned out to be 3158 lb.; curiously, however, while the estimated weight of the Galloways was 1624 lb., the actual weight was 1752 lb., whereas the estimated weight of the crosses was 1437 lb., and the actual weight only 1406. The Galloways are generally considered in the London markets to dress 60 per cent. of their live-weight, but upon this point the editor of the *Herd Book* says that the average weights are as follows:—

	Live Weight.	Dead Weight
Fifteen months	900	540
Twenty-seven months	1400	840
Thirty-nine months	1750	1070
Four years	2000	1240

In 1882 a steer, two and a half years old, weighed over 17 cwt., or 1922 lb., at 1004 days old, showing an average daily gain of 1·9 lb. per day. At the same exhibition a steer, two years eight months and three weeks old, weighed 1754 lb., having gained 1·8 lb. per day. At the Smithfield Show of 1883 a steer weighing 2148 lb. at 1055 days old had gained 2 lb. daily from birth; while at the recent show the heaviest beast, a steer, three years seven months and six days old, which won the first prize, weighed 20 cwt. and 11 lb., showing a daily gain of 1·71 lb. per day; the two-year-old steers at the same meeting reaching in some cases $15\frac{1}{2}$ to $15\frac{3}{4}$ cwt. Although a handsome beast, with a deep well-rounded carcass and fine straight shoulders, the Galloway is not so well finished behind the shoulders, along the back and loins, on the quarters or the flanks, as the Angus; but it is improving, and is perhaps the best-known type of what we may fairly term a mountain breed which exists in any country.

THE AYRSHIRE

To a Scotch dairyman the Ayrshire is almost the perfection of a dairy cow; but even Scotchmen admit that the teats of the udder are too small, especially when men are employed as milkers. Unquestionably the Ayrshire is a very excellent animal. Hardy, less dainty as a feeder than the dairy cows of almost any other British breed, and a deep and reliable milker, she is worthy of all the praise which is universally bestowed upon her by those who have had experience of her work. In colour she varies between red and white, brown and white, yellow and white, and even black and white, the white portion of the coat generally predominating. We may take her average weight at 1000 lb., and her average yield of milk at 550 gals., although enthusiasts have on some occasions declared that good herds average from 750 to 1000 gals., statements which we can scarcely be expected to indorse. The quality of the milk yielded by the Ayrshire is exceptionally good, in spite of the fact that it has been almost systematically described as being more suitable for cheese-making than for butter-making, in consequence of the high percentage of curdy matter, the smaller percentage of fat, and the smaller size of the fat globules. As a matter of fact, the milk of the Ayrshire is rich, certainly containing 12·5 per cent. of total solids, of which, I believe, 3·8 per cent. represents the average of fat. It is not improper to suggest that, if there is any difference, the Ayrshire exceeds the average cow as a fat producer, while it is scarcely entitled to be classed as an exceptionally high producer of solids not fat. Although few Ayrshires are exhibited at the London Dairy Show, they have during three years averaged 44 lb. of milk per day, while the fat percentage is averaged at 4·4, and the solids not fat at 9·4. The fact that selected cows produce such good milk is sufficient to suggest that the Ayrshire is capable of great things; and there is no doubt that by careful selection at least 1 per cent. of total solids, including ·75 per cent. of fat, might be added to the present average production.

The improvement in Ayrshire cattle commenced towards the end of the last century. Ayrshire has long been famous as a dairy district; and the importance attached to milk production had doubtless much to do with the system of crossing and selection, which was carried on with the object of producing a dairy cow of high type. As a rule, while heifer calves are reared, the bull calves are sold and slaughtered soon after birth. The young stock is turned out during the first one or two winters, and its hardy constitution and digestive powers enable it to thrive, the heifers coming into the dairy at an early age in prime store condition. The form of the Ayrshire is somewhat unlike that of every other Scotch breed: the neck and forequarters are exceptionally slender, the carcass gradually increasing in width to the hips, which are placed well apart. The abdomen is capacious, giving plenty of play to the vital organs; and if the hindquarters are not all that Ayrshire breeders would like to see, the udder is more perfectly formed than that of any other dairy cow. The coat is soft and mellow, and the disposition gentle and tractable. In the south-west of Scotland there is no greater treat than to see the large herds, sometimes as many as 250 in number, upon the cheese-making farms of Ayrshire and Wigtownshire. Generally speaking, the bone is fine, while the head is of peculiar form, and adorned with horns which grow upwards, slightly outwards and backwards, coming to a somewhat sharp point.

THE SCOTTISH BREEDS OF CATTLE

According to the standard drawn up for the *Ayrshire Herd Book*, the following are the points of the breed, the points of the bull being somewhat more pronounced than those of the cow :—

1. Head short; forehead wide; nose fine between the muzzle and eyes; muzzle large; eyes full and lively; horns wide set on, inclining upwards 10
2. Neck moderately long, and straight from the head to the top of the shoulder, free from loose skin on the under side, fine at its junction with the head and enlarging symmetrically towards the shoulders 5
3. Forequarters — shoulders sloping; withers fine; chest sufficiently broad and deep to ensure constitution; brisket and whole forequarters light; the cow gradually increasing in depth and width backwards 5
4. Back short and straight; spine well defined, especially at the shoulders; ribs short and arched; body deep at the flanks 10
5. Hindquarters long, broad, and straight; hook-bones wide apart, and not overlaid with fat; thighs deep and broad (but thin of flesh on the inner thigh or twist); tail long, slender, and set on level with the back 8
6. Udder capacious and not fleshy, hinder part broad (and rounded like the side of a cheese), the whole firmly attached to the body; the sole nearly level and extending well forward; milk veins well developed; teats from 2 to 2¼ in. long, equal in thickness, and hanging perpendicularly; distance apart at the sides equal to one-third of the length of the vessel, and across to about one-half the breadth . 33
7. Legs short in proportion to size; bones fine and joints firm . 3
8. Skin soft and elastic, and covered with soft, close woolly hair 5
9. Colour red, of any shade, brown or white, or a mixture of these, each colour being distinctly defined. Brindle, or black and white, is not in favour 3
10. Average live-weight, in full milk, about 10½ cwt. . . 8
11. General appearance and movement stylish 10

Perfection 100

That the Ayrshire is appreciated as a dairy cow need scarcely be remarked. Scotchmen coming south almost invariably bring their cattle with them, if they are engaged in dairy farming. In Sweden and Norway the Ayrshire is also greatly liked, and we have been surprised to find in both countries many herds which are largely composed of the famous Scotch dairy cattle. In the districts of Scotland where Ayrshires are kept, there is a system of leasing the cows, which affords remunerative employment, both to the farmers who own them and to the "bowers" who rent them. These persons pay a specified sum per head for the use of the cow and the food, which is provided by the farmer throughout the entire year. The "bower" is the dairyman—milking the cows, grooming and feeding them, and manufacturing the cheese; and we are bound to say that he does his work well, and keeps his dairy especially clean and smart.

At the London Dairy Show, in 1889, we measured the Champion Ayrshire cow, as follows:—

	Inches.
Height at hips	50
Length	84
Girth round middle	96½
Girth behind shoulder	74½
Across hips	18
Hip-bone to tail	18

CHANNEL ISLANDS CATTLE

By James Long

THE JERSEY

THE Jersey has enjoyed a popularity which has not been exceeded by any other cow among the dairy breeds. It has been alternately lauded and abused—lauded sometimes beyond its merits, and abused by those who have not had a fair experience of the breed. Until recently the Jersey was estimated, even by those who understood it, as a "gentleman's" cow, whatever that may mean; but it has distinctly proved itself to be a strictly economical breed when it is well selected, well fed, and well managed. There is certainly no cow which, under fair conditions, is more profitable to the butter-making or the cheese-producing farmer. Briefly, the economical points of the Jersey are its production of rich milk, rich waxy butter, and exceptionally rich cream and rich cheese. The adjective is applied in each case advisedly, for, with the exception of the Guernsey, which closely approaches it, there is no other variety of cattle in the world which produces either cream, milk, butter, or cheese either so high in colour or so rich in quality. The Jersey is also docile and easily managed, while her appearance always tells in her favour, as it is certainly unapproached; but this cannot perhaps be estimated as an economical property. The points which tell against the Jersey are few, but they are nevertheless important. Specimens of the breed are difficult to fatten, and they never fatten so well or so cheaply as animals of the recognised butchers' breeds, while the bull calves not intended for stock purposes are unsaleable, excepting at extremely low prices as compared with such breeds as the Shorthorn.

It may not, perhaps, be out of place to suggest that when the happy time arrives in which a milk standard is framed by Government, the Jersey will add another point in her favour: her milk will be in large demand as a medium for the enrichment of the average milk of the farm, for without such milk as that produced by the Jersey or the Guernsey a high standard cannot easily be maintained. It is generally admitted that the cream, milk, and butter produced by the Jersey is exceptionally good; but, for some unexplained reason—want of experience, perhaps,—agricultural teachers and writers have been accustomed to speak of the Jersey as ill-adapted for the production of cheese. This is a mistake. I had the advantage of inspecting the three herds in competition at the Chicago Show, where the most important tests ever made, in cheese-making as well as in butter-making, were concluded. The cheese-making results were as follows, and fully bear out my belief that no cow is superior to, while few approach, the Jersey in the cheese-making dairy:—

CHANNEL ISLANDS CATTLE

25 Cows of each Breed tested for 15 Days.

	Milk.	Cheese.	Whey.	Gain. Live-Weight.
Jerseys	13,296·4 lb.	1451·76 lb.	11,578·7 lb.	327 lb.
Guernseys	10,938·6 ,,	1130·62 ,,	9,667·7 ,,	480 ,,
Shorthorns	12,186·7 ,,	1077·6 ,,	10,838·9 ,,	709 ,,

No cows have excelled those which have been produced from Island and English bred cows in America. In some cases as much as 900 lb. of butter have been produced from a single cow in 365 days; while in many instances, to borrow a paragraph from my own work, *The Elements of Dairy Farming*, bulls of particular strains have produced heifers through dams, daughters of their own, which have given record yields of butter, as follows:—

Purest, 15 lb. 4 oz. butter per week. { Mercury, 432 { Jupiter, 93 { Saturn, 94 / Rhea, 166 / Alphæ, 171 { Saturn, 94 / Rhea, 166 / Nymphea. 18 lb. 7 oz. { Mercury, 432 / Phedra. 19 lb. 12 oz. { Mercury, 432 / Leda, 779 14 lb. { Jupiter, 93 / Europa, 176 15 lb. 6 oz. { Jupiter / Alphæ, 15 lb. 8 oz.

The Jersey is often held to be a delicate cow, but this is not the case; she succeeds in Scotland, and in one of the coldest parts of Canada I have seen her thriving as well as in Jersey itself. The Island, however, is the place to see the Jersey in its glory: she is smaller, it is true, but she is more slender and beautifully proportioned; and no more perfect example of the breed could be imagined than such an animal as the famous "Rosy," which was well known in our show-yards, and which was bred in the Island. What can be done by the Jersey is shown by the fact that in Mr. Dauntsey's herd fifty cows averaged a return of £32 per head in a single year; while, coming to a more recent date, the herd of Mr. J. B. Ellis, a well-known and extensive Norfolk farmer, has returned him a sum which, in these days of agricultural depression, is, if not quite so large, sufficient to warrant all the praise that is bestowed upon the breed.

The Jersey varies in colour, from golden-fawn to mulberry, while she embraces a number of shades which are affected more or less by the introduction of grey; practically, the colours are fawn, golden, golden-fawn, silver-grey, and mulberry. Sometimes these colours are splashed or broken with white. The weight of an average cow is about 830 lb. The average yield of milk is about 450 gals., but this quantity is easily exceeded by careful selection and high feeding. To be just, however, as well as generous, to the breed, we are bound, in giving an average, to embrace the inferior as well as the superior cattle; similarly, the average quality of the milk is about as follows: fat, 4·64 per cent., solids not fat, 9·32 per cent. An examination, however, of the records of the trials and tests at the Dairy Show and elsewhere will show that occasionally the quality falls below these figures, although perhaps it is more often higher. The Jersey possesses a delicate thin skin of extremely soft texture; and in no part is it more mellow or silky than upon the udder, which is improving in form, for it is not quite so globular as it might be. The cow is not a well-fleshed animal. While bright and healthy in appearance, the skin is apparently drawn over a frame scantily fleshed, but beautifully proportioned. Naturally, the majority of the cows are far from perfect, but the imperfection is generally in the hindquarters, the tail being set on too high, while the hips are not wide enough apart, nor do the buttocks permit of sufficient expansion of the udder. The Jersey has a black tongue, black nose, and black points generally; while her skin is yellow, sometimes a rich orange, and somewhat oily withal. The horns are crumpled, or growing forward and then curved inwards. In very choice animals they are extremely fine, yellow at the base and black towards the point. The forequarters are slender, the shoulder being almost imperceptible; the ribs are fairly well sprung, the back straight, and the loins deep; but there is a cavity between the forepart of the udder and the base of the abdomen which breeders, as a rule, would like to fill up with a more extensive milk vessel. To give an idea of a type of a first-rate cow, we may take "Baron's Progress," the animal which won the champion prize in the butter test at Islington in 1889, producing in a single day 3 lb. 5 oz. of butter. We took her measurements, which were as follows:—

Height at the hips	50 inches.
Length	87 "
Girth round shoulder	72 "
Girth round middle	91 "
Across hips	16 "
Hip to tail	20 "

What the Jersey has done at the milking trials of the British Dairy Farmers' Association has been shown by Mr. J. F. Hall, an enthusiastic admirer of the breed, in the following figures. These show that Jerseys milk well, both as regards quantity and quality, to a considerable age:—

Aged.	No. tested.	Days in Milk.	Average Milk.		Average Fat.	Average Solids other than Fat.	
			lb.	oz.	oz.	lb.	oz.
2– 3	29	65	24	0	10	1	2½
3– 4	8	67	26	6	11	1	4
4– 5	12	75	29	11	12	1	6½
5– 6	13	87	29	11	12¼	1	6
6– 7	9	93	29	13	14¼	1	6¼
7– 8	6	85	32	0	12¼	1	8½
8– 9	7	99	28	6¼	12¼	1	6
9–10	2	103	28	14½	11	1	6¼
10–11	2	90	31	9	14½	1	7½

The Jersey is supported by a society known as the English Jersey Cattle Society, the secretary of which is Mr. John Thornton, of 7 Princes Street, Hanover Square. In the Island is another society which issues a Herd Book, the secretary of which is Mr. Joshua Legros, of St. Heliers.

THE GUERNSEY

Some figures, which have been given above for comparison with the Jersey, are sufficient to show that Guernsey cattle have not only immensely improved during the past few years, but that they are amongst the most prolific and economical cattle in use for dairy purposes. At the same time, these figures do not entirely represent the actual capacity of the breed. The collection at Chicago, good as it was, was not what it might have been; and the same remark applies to many of the collections which are seen at English shows, few of which can compare, as milking cattle, with animals which are seldom seen in public. As a matter of fact, the show Guernsey, like the show Jersey, is not always the best milking animal. At

Chicago the twenty-five selected cows were not the best America could produce, inasmuch as their owners had refused to part with their choicest cattle; but they were fairly representative of good animals of the breed. We had the advantage of inspecting the best large collection in the States, that of ex-Vice-President Morton, who is now the Governor of New York State, and who, as a millionaire, keeps up his Guernsey dairy farm on the Hudson River regardless of cost. This herd is very fine, and splendidly managed; but no single member of the herd could compare with some of the best cattle which we remember to have seen at the Island Show; indeed, Guernsey cows are there exhibited which, as milkers, are probably superior to anything which has been in competition for milking prizes in any English showyard. That Guernseys with great capacity can be purchased in the Island for economical purposes at moderate prices cannot be too well known. On one occasion we were shown a herd, owned by Mr. Christopher Middleton, among which were cows which had exceeded a thousand gallons in the year, and this is saying a great deal for an animal of moderate size.

The Guernsey is an animal of a bulkier and, shall we say, of a more massive type than the Jersey. Her colour is orange, or orange and white. In weight she averages about a 1000 lb. To be exact, we may quote the following figures, taken from the Dairy Show records for four years, comparing the Jersey with the Guernsey, from which it will be seen that we are in neither case wide of the mark in the weights which we had taken and published before this record appeared :—

	1890.	1892.	1893.	1894.
Guernseys	1026 lb.	1044 lb.	1046 lb.	1136 lb.
Jerseys	840 ,,	829 ,,	824 ,,	842 ,,

The weights of 1894, which were unusually heavy in the Guernsey class, are based only upon a couple of entries. The average yield of milk of the Guernsey was placed at 520 gallons, while the average quality may be suggested as : fat, 4·55 per cent. ; solids not fat, 9·25. In 1892 and 1893 the Guernseys averaged 4·91 per cent. of fat, and 9·04 per cent. and 9·22 per cent. of solids not fat respectively. In 1895 the fat was a little higher, and the other solids a little lower. The quality of the milk, however, varies, and a strict average can only be based upon a large number of tests. Sometimes the Guernseys reach the highest figures attained by the Jerseys, and sometimes they fall very nearly to the average of ordinary cattle. The Guernsey, however, is a breed which is capable of being developed in a high degree, and it is quite on the cards that, in future, averages from 800 to 1000 gallons may be obtained over a considerable number of cows in a single herd, while maintaining the high quality of the milk.

The points for and against the Guernsey are practically the same as those we have enumerated in connection with the Jersey. The milk, cream, butter, and cheese are all exceptionally rich, while the colour is high. The cow is also extremely docile, and presents a very handsome appearance, although we may be pardoned for saying that she does not equal the more fawn-like animal, either in form or colour. Like the Jersey too, the Guernsey cow is not easily fattened, nor are the calves so saleable, but both Jersey and Guernsey calves may be converted into veal if taken early in hand and allowed all the milk they will consume until they are ready for the butcher. Occasionally steers have been exhibited which have been creditably fattened; but if the cost of the production of the meat they carry were exactly estimated, we are afraid that the process of fattening would not be considered very economical as compared with some other breeds.

The Guernsey has a white nose and tongue, light-coloured horns, a heavier front than the Jersey, a soft skin, somewhat greater width across the loins, well-sprung ribs, and depth of both carcass and hindquarters. We measured a champion two-and-a-half-years old heifer at the Dairy Show as follows:—

Height at the hips	50 inches.
Length	86 ,,
Girth round the middle	91 ,,
Girth at the shoulder	72 ,,
Across the hips	18 ,,
Hip to tail	18 ,,

The Guernsey Society has done good work in bringing the breed to the front, and in offering prizes for butter tests similar to those which have been successfully carried out by the Jersey Society. The Secretary is Mr. Julian Stephens, 19A Colman Street, London, E.C.

THE KERRY AND DEXTER-KERRY BREEDS OF CATTLE

By John Watson

"There is no doubt this is distinctly the poor man's cow, thriving where no English or Channel Islands breed could get a living, and during extremes of heat and cold is of the hardiest possible constitution. It has beaten all other breeds in the question of milk in proportion to live weight, and, as the Prince of Wales remarked some years ago, produces the finest beef in the world."

This is the opinion of a disinterested authority on Kerries, whose knowledge of the breed is second to none; and on its practical side it is strikingly borne out by the following facts:—"Red Rose," a seven-year-old Dexter cow, and a good type of her breed, gained the first prize in her class at the "Royal"; she calved at the end of March in one year, and during the following twelvemonth gave a supply of milk which amounted in the aggregate to a weight of 4 tons 9 cwt. 3 qrs. 20 lb., or, measured in the usual way, 983 gals. "Red Rose" is a small cow, as all her breed are, and herself weighs just 9 cwt.; so that during the year she produced more than ten times

THE KERRY AND DEXTER-KERRY BREEDS 53

her own weight. Another cow, also a prize-winner, was, when I saw her, giving 16 quarts daily—a capital return for a small cow; and this, too, in December.

Although this is perhaps an extreme case, it is by no means so exceptional as it would seem, especially among the best examples of the breed. For instance, the writer recently saw a small red cow, the property of Mr. Martin J. Sutton, and winner of the first prize at the "Royal," which gave her own weight in milk in seventeen days. The butter from the milk of the Kerry is of very high quality, and perhaps the best testimonial to the breed in this connection is that it is now being adopted as the stock animal of several professional dairy farms. As a rule, the cows have remarkably large and well-shaped udders.

That Kerries make good beef is now amply demonstrated, and a splendid steer exhibited at Smithfield fetched £40, killing remarkably well. The beef is fine in grain and rich in flavour. And just now, when there is such a run in favour of small carcasses, no variety of cattle when fat sells more readily, or at a higher price in relation to weight, than the Kerry and Dexter-Kerry. Without here going into the question of the difference in the two breeds, it may be briefly stated that the strong point of the former is its milk-producing qualities; and of the latter, meat and milk producing properties combined. Of the two varieties, however, more hereafter.

There is no thriftier animal in existence than the

Kerry; and although it manages to thrive on its bare native hills, it soon responds to better food and conditions when it comes south; for, like its fellow-countrymen, it makes a capital emigrant. It has travelled south with every indication that it has come to stay, and it is only to be hoped that its usefulness and thrifty nature will not be spoiled by the over-generous treatment it is now receiving. For farmers, the practical outcome of pampering is that it would lose two chief characteristics—its good milking properties and its hardihood. Some of the chief English herds, however (and, curiously enough, the best breeds of Kerries are now found in England), are running on poor light land, and, beyond hay, get little in the nature of special food. An advantage of the Kerry is that three of them can be kept in place of two cattle of any other breed, and they grow fat on land that would starve a Hereford or a Shorthorn.

In one connection the writer feels confident in prophesying the ascendency of the Kerry, namely, where one or two cows are kept for supplying a family with milk, or as ornaments in a home paddock. The Jersey or Guernsey usually occupies this position; but, in addition to the care and careful housing these breeds require, the percentage of losses is great—so much so that in the north of England and Scotland the Channel Islands breeds are practically out of court. In the more hilly parts of the country it is a question whether the Kerry would not prove a more paying animal than the poorer class of Shorthorns which now pasture there. As an admirer of the latter breed (living, as I do, in the heart of the Shorthorn country), I believe it to be by far the best breed of British cattle; but when I see it in the hill-country valleys, with good frame, but otherwise a shadow of what it ought to be,—in such cases I am confident the Kerry would prove the better animal. In making this suggestion, I do not mean that it should invade the great belt of pasture occupied by the Ayrshire, an all-round animal which cannot be excelled in its native country.

It is interesting to note that the Kerry has been largely employed in carrying out the plan of renting cows to labourers and cottagers on the hiring-out system—another proof that it is essentially the poor man's cow. Individual cows, originally imported as in-calf heifers at from £7 to £8, have, even after having been fed for years, been sold at a profit. When anyone sees a Kerry for the first time, astonishment is often expressed at the small size of the animal; and it is said that the Royal Dublin Society awarded the prize to a cow measuring only 38 in. at the shoulder, 70 in. in girth, and 42 in. from top of shoulder to setting on of tail.

As its name implies, the Kerry has its original home in Ireland, and it is the most appropriate breed of cattle to the bleak Kerry hills which it pastures. How long it has been in Ireland is not known, and many different accounts are given of its origin. As long ago, however, as a hundred years its merits were known; for when, on a large estate in the county of Cork, an experiment was tried, the Kerry came out best from among all the breeds for the quantity of good milk it yielded. And early in the present century, as the outcome of a survey undertaken by the Royal Dublin Society, these cows are praised as good milkers, one weighing 3 cwt. yielding not uncommonly 16 quarts a day.

In that day the Kerry farmers preferred small beasts, as peculiarly capable of enduring hardship and being easily maintained. In fact, it is the small size and the abundant yield of the Kerry that renders it of such value to the small farmers and cotters of Ireland. The keen interest now taken in the breed dates from 1887, when a Herd Book was established to promote its interests. It is described originally as a light, neat, active animal, with fine and rather long limbs; narrow rump; fine, small head; lively, projecting eye, full of fire and animation; with a fine white cocked horn, tipped with black; and in colour either black or red.

Another interesting fact in connection with the development of the breed is its recognition by the show authorities. Kerries were first exhibited at the Royal Dublin Society's Show in 1844, but not until 1876 was a distinct class provided for Dexters. Kerries were first provided with a class at the "Royal" in 1862, while by 1889 (at Windsor) the two breeds had so gained in popularity that 136 animals were exhibited. Another important step was when the Smithfield Club provided a section for Kerries and Dexter-Kerries, at the same time establishing classes for small breeds. The importance of this step is that by so doing these cattle were for the first time recognised as beef-producers. In this connection, and

THE KERRY AND DEXTER-KERRY BREEDS 57

in confirmation of the opinion of H.R.H. the Prince of Wales, quoted at the beginning of this chapter, it may be mentioned that after the Smithfield Show of 1890 a three-year-old Kerry heifer was followed to the butcher, where it was stated that a more useful or satisfactory carcass had never been seen, the lean meat being juicy, of fine colour, and full of "nature." The live-weight was 10 cwt. 1 qr. 26 lb., and the dressed carcass weighed 726 lb., showing a percentage of dressed carcass to live-weight of 61·84. The forequarters weighed 23 st. 4 lb., and the hindquarters 21 st. 7 lb. Kerries are now more and more in request for crossing with larger breeds (such, for instance, as the Aberdeen-Angus) for the production of "small beef." Whether at home or abroad, Kerries have always fattened well, and the beef is nicely marbled. In fact, the Kerry is, both as a milker and as a beef-producer, equal, if not superior, to any other breed of British cattle.

An important variety of the Kerry is the Dexter-Kerry, an animal that differs from the parent stock mainly in roundness of form and shortness of leg; it is also larger. Whether this variety was produced by selection, or whether size has been obtained by crossing with some larger breed, it is impossible to say. Suffice it that an animal has been produced feeding up to a wonderful depth and thickness, on a compact frame; and it has been said of a Dexter-Kerry that it presents the appearance of a grand Shorthorn seen through the wrong end of a telescope. Whatever else the Dexter may be, it is an improved "Kerry," and the quotation which follows conveys a practical description of this black diamond. It is "a Kerry pressed down, flattened, moulded, smoothed, and widened into a plump, dumpty little beast, with a softer and somewhat larger head, and short, straight posts of legs, which give her a dwarf-like appearance. With the exception of one or two points, the transformation is complete; in fact, there is nothing now left but the horns to identify her with her original family. To sum up, the Kerry may be described as a picturesque, hardy, mountain goat-cow; the Dexter as a quaint-looking little domestic cream-ewer; and both breeds as the *crème de la crème* of the dairy, and *bonne bouche* of our 'English roast beef.'"

The following are the standard descriptions of the Kerry and Dexter-Kerry:—The Kerry cow should be long, level, and deep; her colour black; her head long and fine; her horn fine at base, mottled or white, tipped with black, upright and cocked; her eye soft and prominent; her bone fine; her coat like satin in summer, in winter long and thick; her udder should be soft and large, but not fleshy, protruding well under the belly, the teats being placed square and well apart, the milk veins prominent and large; the tail should be well put on, and have at the end long fine black hair. The Kerry cow should not weigh over 900 lb. live-weight when in breeding condition. A small amount of white on the udder and under line not to disqualify. The bull should be whole black, without a white hair; should have a long head, wide between the eyes, of masculine character; throat clean; horns medium length, mottled or white, with black tips, turned backwards; withers fine; back straight from withers

THE KERRY AND DEXTER-KERRY BREEDS

to setting-on of tail, which should be long, fine, tipped with black hairs. The Kerry bull should not weigh over 1000 lb. live-weight when in breeding condition.

The Dexter is essentially both a milk-producing and a beef-making breed, and both these points should, in judging, be taken into consideration. Bulls, whole black or whole red (the two colours being of equal merit); black with white on udder, or red with white on bag. The extension of the white of the udder slightly along the inside of flank or underside of belly, or a little white on end of tail, shall not be held to disqualify an animal which answers all other essentials of this standard description :—Head short and broad, with great width between eyes, and tapering gracefully towards muzzle, which should be large, with wide-distended nostrils. Eyes bright, prominent, and of a kind and placid expression. Neck short, deep, and thick, and well set into the shoulders, which, when viewed in front, should be wide, showing thickness through the heart, the breast coming well forward. The horns should be short and moderately thick, springing well from the head with an inward and slightly upward curve. Shoulders of medium thickness, full and well filled-in behind; hips wide; quarters thick and deep; back straight; ribs deep and well sprung, flat and wide across loins, well ribbed-up, straight underline; udder well forward and broad behind, with well-placed teats of moderate size; legs short (especially from knee to fetlock), strong and well placed under body, which should be as close to the ground as possible; tail well set-on, and level with back. The skin should be soft and mellow, and handle well; hide not too thin; hair fine, plentiful, and silky. Dexter bulls should not exceed 900 lb. live-weight when in breeding condition. Dexter cows should not exceed 800 lb. live-weight when in breeding condition.

DEVON AND SUSSEX CATTLE

By James Long

An indication of the growing popularity of the Devon breed of cattle is to be found in the fact that fifteen classes were provided at the great Chicago Exhibition for this grand beef and milk producing variety of British stock. Special provision was made for three distinct sub-varieties of the breed, namely, those of the *American Devon Record*, Davy's *Devon Herd Book*, and the *Canadian Devon Herd Book*. This fact shows that Devons have travelled far afield, and that the breed may now be considered cosmopolitan in its distribution. The three sub-varieties named above were shown in no less than fifteen sections, perhaps the most generally interesting being those in which the animals competed in herds of aged and young animals respectively. The first was composed of one bull of two years old and under three, one cow three years old and under four, one heifer two years old and under three, one heifer one year old and under two, and one heifer under one year old. The young herd consisted of one bull and four heifers, all under two years old, and bred by the exhibitor. Two other interesting sections consisted of four animals of either sex, under four years old, the get of one sire; and of a cow or two of her produce.

These sections have been set out at some little length as, in addition to all the ordinary classes, they represent the treatment which the Chicago Agricultural Commission accorded to more important English breeds.

Originally, Devon and Sussex cattle were peculiar to the south-eastern and south-western districts of England, in which the counties from which they take their name are situated. Speaking generally, they can be said to possess the high quality of the Shorthorn, the Hereford, or the Angus, but they are hardly such economical feeders, and unless in exceptional herds, where they have been specially bred and selected for the purpose, they are not the best of milkers. They are the two essentially whole-coloured, red, horned breeds of Great Britain, and yet they are unlike in character, —the Sussex reminding us of the draught cattle of old times; indeed, they can even now be found upon the Southdowns ploughing up the land, and driven by a goad, although upon what principle of economy they are preferred to horses we have been unable to discover. The Devon—smaller than the Sussex, yet more compact and symmetrical, and carrying meat of finer quality—is a grazing beast admirably adapted to summer feeding and to the requirements of a great grass district like the south-west of England. It is a peculiarity of farming, whether it is applied to England or to Continental countries, that local requirements or local sentiments have developed a variety of breeds of cattle which are more or less meritorious, but which are maintained with all the ardour which breeders infuse into a subject closely connected with the county or province in which they were cradled, and in which they will probably die. We see the Shorthorn and its half-bred progeny everywhere, even in the strongholds of other breeds. The Hereford and the Scot are occasionally met with in all parts of the country, but the Sussex is rarely seen outside the county. The Devon, however, is perhaps more popular,—many southern and western farmers selecting it for grazing and winter feeding,— but it is practically confined to the south-western district of England.

THE DEVON

Devon cattle may be divided into three classes, the North Devon, the South Devon or South Ham, and the Somerset Devon, all having a wider distribution than their names suggest. The last named is the most popular, and maintains the prestige of the breed at Smithfield and other exhibitions. If smaller than the larger beef breeds, it is equally symmetrical, and its breeders claim that its flesh is better marbled and firmer, of finer texture and generally superior quality than that of any other pure breed. Certain it is that this statement has actually been borne out when, as in 1891, a Devon won the London championship, excellent practical judges, who subsequently examined its carcass, declaring that they had never seen anything superior. Although by no means diminutive, the Devon is not a large beast, and cannot compare in size with the massive Hereford or Shorthorn, or even with the Sussex, Polled Scot, or Welsh breeds. It is really a breed of medium size, and yet its kinsmen of the north of the county are smaller, although they are sometimes used in crossing; while the South Devons, which are bred along the coast, are larger, bulkier, and coarser. The standard colour is red, and it is very solid, although in Devonshire white marking is frequently found associated with it. The nose is white, the horns prominent, but fine in texture, and gracefully spread in an outward and upward direction, and they taper gently to a point. Compared with the bones of the Devon, the carcass is large; and this point may be specially noted, as it is one which is characteristic of all first-class meat-producing cattle.

It is probable that no breed of cattle produces more meat in proportion to its size, or in proportion to the food it consumes; and it may be added that there is no ox which carries less offal to its carcass. The fattening process is rapid in the stall, and almost equally so upon the pasture, to which it is specially adapted. Upon Dartmoor and Exmoor, where a rougher class of Devons graze, they thrive excellently, withstanding the rough weather of these exposed districts as well as the Ayrshires, the North Welsh, and the Kerry withstand exposure upon their respective native mountains.

If we look at the Devon as a dairy cow we shall not find her wanting. Here again, however, there are Devons and Devons. The cows of a selected herd produce milk of an immensely rich quality, if it is smaller in quantity than that produced by the Shorthorn and the Ayrshire. We have met with numerous instances in which the milk from a Devon herd has equalled that of an ordinary herd of Jerseys, which is saying a good deal; hence it is not uncommon for Devon cows to produce a pound of butter a day. The Devon is largely used in the great cheese districts of Somerset and Dorset, and the richness of its milk has something to do with the high quality of the famous cheddar cheese. Unfortunately, the system of recording the daily yield of milk in each cow is not very common in the west of England. Were it recognised by owners of Devon dairy herds, as it is by owners of Jerseys and Guernseys, and in some counties of Shorthorns, we should be in a position to say more of the milking qualifications of the Devon than is possible under existing circumstances.

It has been pointed out that the Devon has in parts of the west of England pushed out the Shorthorn, which was formerly the established breed; and that at the Government Farm at Dartmoor, where it was originally kept, after giving way to the Ayrshire and the Polled Scot, it has once more been established as the presiding breed. One of the largest and best breeders of Devons told the writer, some years ago, that he could keep three Devons to two Shorthorns, and that he finds they withstand the winter and the wet climate, without extra housing or feeding, better than any other pure breed or cross breed. This gentleman feeds a large number of cattle, summer and winter; and if, as sometimes happens, he gets hold of a Shorthorn or Hereford, or a cross-bred beast, he usually finds that they do him no service.

With regard to weight, we take the figures furnished to us by another well-known breeder. He says that the live-weight of a "Somerset" Devon reaches to from 18 to 22 cwt. at four years old. Steers under two years old he puts at from 8 to 10 cwt., steers under three at from 12 to 15 cwt., and under four at from 13 to 17 cwt., while the cows vary between 12 and 17 cwt. We cannot, however, do better than quote, from published data, figures referring to prize and other animals exhibited at the great Smithfield Show of December 1891, adding some figures obtained by the late Mr. G. T. Turner for the *Live Stock Journal*:—

	Age in Days.	Live-Weight in lb.	Average Daily Gain in lb.	Weight of Dressed Carcass in lb.	Per cent. Carcass to Live-Weight.	Weight of Bones and Fat in lb.	Weight of Hide in lb.
Steer, 1st .	558	964	1·73	642	66·60	85	64
,, 2nd	1014	1430	1·41	988	69·09	...	99
Champion .	954	1556	1·63	1072	68·89	...	...
Ox, 2nd .	1448	1655	1·14	1120	67·67	...	80
,, 3rd .	1386	1672	1·22	968	57·89	...	84
,, com. .	1309	1746	1·33	1120	64·14	...	...
Heifer, 1st	880	1346	1·53	856	63·60	...	...

These figures compare exceedingly well with those compiled at the same time to represent other breeds. The high percentage of carcass to live-weight was only exceeded in a few instances among the numerous animals of which particulars were forthcoming; indeed, the two-year-old steer beat all other breeds but the Shorthorn. As much, however, cannot be said of the weights which were reached—the Devons, with one exception, weighing less than any of the other animals tabulated; but it was in the daily average gain of live-weight that the Devon showed to the worst advantage. Whereas two-year-olds of the large and cross-breeds frequently exceeded two pounds per day, the Devon, as shown above, reached only 1·73 lb.; while the three-year-olds,—oxen and heifers,—including the champion of the show, gave still lower weights. It should, however, be mentioned that the champion Devon steer was said by Mr. Turner to have the least wasteful carcass of any animal he ever saw. This animal was slaughtered by Jewish butchers; the bone was very fine and the meat grandly marbled. In some other cases fault was found by the butchers on account of the excessive quantity of fat, but the quality and flavour of the meat seems to have been highly praised. Take it all in all, the Devon of the showyards is hard to beat.

Reviewing this breed of cattle for the year 1894, a writer who is in close touch with Devons remarks as follows upon their beef-producing capacity:—"It is readily admitted by the most enthusiastic admirers of the 'Rubies' that a 14 cwt. Devon will not produce as much as an 18 cwt. or 19 cwt. ox of some coarser breed; but the difference from a pecuniary point of view is often on the side of the smaller animal from the west, when the amount of food consumed is taken into consideration; whilst it is well known that the choice flesh of the Devon, which is of such remarkably fine texture, is put on with amazing rapidity." Taking into account, therefore, their robust constitution, smallness of offal, gentle temper, great symmetry, aptitude to fatten at small cost, and ability to put on flesh on the right parts of the carcass, Youatt was not far wrong when, alluding to the merits of this breed, he said: "The best of them are the best in the world." Since that expression was employed, the beautiful red cattle have improved instead of deteriorated; so I feel confident that they will continue to hold their own with any other breed in Great Britain or elsewhere; whilst the possession of two champion awards at the Smithfield Show within the space of four years furnishes ample proof that Devons still occupy the high position they have hitherto held in the cattle kingdom.

THE SUSSEX

Sussex cattle have undoubtedly improved during the last dozen years, especially in the hands of those breeders who are represented by the *Sussex Herd Book*. They are considered to be rapid beef-makers but inferior milkers, and we shall see from established facts how far this is true. Some years ago the writer of these lines prepared a report upon the herds of British cattle for the United States Government, and various opportunities were consequently afforded of obtaining data, which was of an interesting as well as instructive nature. To commence with, it was admitted that Sussex cows seldom gave sufficient milk to meet the requirements of their own calves. Now it is a well-known fact that some breeders, leading exhibitors of stock, make a practice of keeping calves upon their foster-mothers until they are a year old. The Sussex cow can hardly be equal to this. If we look upon the Sussex as a useful and still improving beast, we shall do it justice; to say more would not. It cannot be classed with either the Shorthorn, the Hereford, the Scot, or the Devon; it comes, indeed, in the second category of economical British breeds—with the Welsh and the Red Poll, both of which are, like itself, improving. We remember very well when the Sussex was raw, leggy, and entirely wanting in that form which is recognised as indicating character and quality in a beef-producer. Now, however, it is being licked into shape by ardent and enterprising men, who are anxious to see it resemble the breeds we have named. At this moment, by comparison, the body is not sufficiently rectangular: there is not enough depth and width of loin in comparison with the size of the animal; the bone is large, and the hide thicker and coarser than it ought to be; nor is the Sussex so kind a feeder, producing the mellow flesh of the best breeds and their crosses. When, however, we compare it with its peers of the second grade, with those which are its inferiors, or with continental breeds of cattle, we find it holding a high position. One of the best Sussex breeders furnished the following measurements and weights of London prize animals:—

Heifer, 3 years old . . . 7 ft. 8 in. by 4 ft. 9 in.
55 score or 1100 lb. Live-weight, 1708 lb.
„ 1 year 11 months old . 7 ft. 7 in. by 4 ft. 8 in.
52 score or 1040 lb. Live-weight, 1652 lb.
Steer, 1 year 11 months old . 7 ft. 4 in. by 4 ft. 4 in.
46 score or 920 lb. Live-weight, 1428 lb.

That great live-weights have been attained is perfectly well known; but the larger bones, and sometimes the heavy intestines, account for it in the live beast. It sometimes happens that the daily gain of live-weight and the percentage of carcass weight to live-weight are both excellent, showing that the gain in weight is not altogether due to the valueless part of the animal. Let us examine the particulars furnished regarding this breed:—

	Age in Days.	Live-Weight in lb.	Average Daily Gain in lb.	Weight of Dressed Carcass in lb.	Per cent. Carcass to Live-Weight.
Steer, 2nd	679	1571	2·31	995	63·34
,, (cup)	1030	1883	1·78	1232	67·21
,, 3rd	915	1856	2·03	1278	68·66
,, com.	970	1770	1·82	1188	67·68
,, 1st	1275	2049	1·61	1410	68·81
,, 3rd	1178	2104	1·79	1440	68·44
Heifer, 1st	1023	1689	1·65	1168	69·24

The above figures show in several cases excellent results, although some animals came out much better than others. The high daily gain of the yearling is spoiled in value by the low carcass percentage. Taking both features, the 3rd prize two-year-old steer comes out second only to the Champion Shorthorn, the best butchers' Scot and cross bred. The figures, as thus taken, are very remarkable for the tale they tell of the judging. The handsomest beast in the ring frequently takes a very back seat when he reaches the block; and, after all, it is "handsome is as handsome does." As a rule, an ox of correct type, well fattened, will produce better results than one of inferior type, however well he may have been fed, for he carries most meat on the best parts, and there it has much the greatest value. Mr. Turner examined a number of Sussex carcasses at the Christmas Show of 1894, and he quotes, in the report already referred to, the remarks of the butchers who purchased them. In most cases the meat was described as of good quality—the quantity of loose fat being small; but in one instance the results were not so satisfactory, the fat over the loin and ribs being four inches thick.

The Sussex is a red beast entirely whole-coloured; his coat is long and fine; his horn is of a medium length, and grown forward and slightly upward. He has a good round barrel, with broad ribs; a tolerably flat loin and straight back. The rump in good cattle is long and flat, with plenty of width where the tail is set on. Compared with the Devon, the legs are long. The breeder does not take his first calf from his heifers, as a rule, until they are three years old, which adds materially to their cost. With some milking cattle it is not uncommon to have a second calf by the time the same age is reached, having already obtained a season's milk. Latter-day experience suggests that early maturity applies alike to breeding as to butcher's stock. Sussex steers are still put to the draught yoke, where this plan is continued, from the age of three until six. They are, however, slow, and for some time do not get through much work. As three-year-olds are brought into work, six-year-olds are turned out, and either sold or fed off for the butcher. Oxen of the same age are kept together, as the older are stronger than the younger ones, and bear the labour of their burden better. A Sussex team is still a sight to see, carrying the farmer of to-day back to old days, in which alone the majority have seen them at work, or to foreign countries, where the draught ox is much more commonly worked than with us.

HEAVY HORSES: BREEDING AND MANAGEMENT

By Gilbert Murray

During the early days of railway enterprise the opinion prevailed that the demand for heavy horses would decline, through the displacement of the slow and cumbrous road waggon of a remote date; and a wail was set up that horse-breeding was doomed. So far from this being the case, the demand has steadily increased, until at the present moment the demand for the best class of heavy horses for town work has exceeded the supply, and prices have increased 50 per cent. Some forty years ago sound geldings, fit for the general purposes of agriculture, could be purchased at any large fair in the Midland Counties at from £20 to £30 each—the best horses that could be met with would not fetch more than £40. Now the best class of heavy horses, rising six years, with action, sound, and good workers, readily sell at from £80 to £90 each. During the forties and early fifties the rage for two-horse ploughs was at its height in the south, and the sluggish, sturdy-legged horses of the Midlands were on most of the light lands superseded by the more active, clean-legged Yorkshire or Cleveland type. Long prior to this date the cultivation and improvement of the Clydesdales—a breed indigenous to the best cultivated districts of the west and south-west Lowlands of Scotland—had been commenced. These were much improved by the importation of a heavier class of mares from Lincolnshire and the Midland counties. During the last fifteen years the establishment of Stud Books, and the enthusiasm devoted by rich men to the improvement of heavy horses, has been of immense benefit to the tenant farmer, by placing within his reach the use of well-bred sires at a moderate fee. Take the county of Derby as an illustration. There the proportion of tillage is limited to that of grass, hence there has been less need for the substitution of a more active race. When the reaction again set in, here was a sound foundation, which, though not registered, could be traced by oral tradition through many generations. Many of the owners of these mares and their descendants have reaped rich harvests from the sale of their produce, which has now been widely distributed. Many of the best and purest strains are getting into the hands of rich breeders, who form studs for stallion-breeding and animals for show purposes. Nevertheless, the ordinary farmer who has a well-bred mare, and who mates her with a suitable stallion, may occasionally breed a valuable animal. The breeder should be particularly careful in guarding against any semblance of unsoundness whether in the sire or dam; then, should he fail to breed a prize-winner, he is, under ordinary circumstances, pretty certain to produce a serviceable animal for general purposes.

Our object is, however, to guide the tenant farmer

HEAVY HORSES: BREEDING AND MANAGEMENT

in the selection and breeding of commercial animals, and, as far as possible, to assist horse breeders over the present depressed period, rather than the rich amateur, whose object is chiefly to assist his tenants and dependants. In the case of the farmer, breeding from young mares has much to recommend it. With him time is an important factor. The mare, if well-cared for, should be put to the stud at two years old. She should be broken in and worked at the plough during the winter. As soon as she has foaled, and there is a sufficient bite in the pastures, mare and foal may be turned out and allowed to remain till weaning, at the end of September or early in October. When the foal is weaned, the mare should have a dose or two of aperient medicine, and the milk should be drawn by hand for the first three or four days, when she will be able to take her place in the team. Early breeding may be safely practised without the slightest danger of deterioration or stunting, if the mare be allowed to remain idle whilst nursing her foal. No greater error can be committed than that of working a mare during the time she is suckling a foal. The former is weakened and emaciated, whilst the latter is starved and stunted. The most successful breeders are those who use matured and comparatively aged stallions on young mares. The reason for this course is obvious. When a stallion is used in a district for seven or eight years, the observant breeder is enabled to form an idea as to the value of his stock and the class of mares with which he most successfully "nicks." As a rule, the most successful mares for breeding the best and most valuable class of horses for heavy work are short-legged animals, which look to be 15 hands, and yet measure 16 hands when placed under the standard. The ribs should be well sprung, and the legs placed outside the body; hocks clean and broad when examined from the side, and slightly inclined inwards when seen from behind; the forelegs should be short from knee to fetlock, bone flat, and supported by strong muscles. The breeding mare, when in the hands of a steady, good-tempered man, may be worked regularly to the date of foaling; in fact, steady work is conducive to the health both of the dam and her offspring.

Within the last decade much more care and attention have been devoted to the feeding and management of the draught horse than formerly. The food is now generally prepared, and the quality and quantity regulated in some degree with reference to the requirements of the animals. This, again, depends on the amount of work they have to do. At no very remote date it was a standing rule to allow each farm horse two bushels of oats per week, or 12 lb. per day—fed without any preparation,—and, in addition, an unlimited quantity of hay or straw. This, and an evening mess of boiled turnips and chaff, was the daily allowance during the winter months, the long days, and busy period of turnip sowing. Owing to the small comparative capacity of the stomach of the horse as compared with the bulk of the animal, concentrated food is of the utmost importance. Large quantities of long hay and straw are positively injurious, even dangerous, from their liability to cause the troublesome disease known as impaction of the *rumen*. When the horse returns to the stable after a hard day's work, in a heated, exhausted, and hungry state, as soon as the harness is removed a feed of unprepared oats is placed before him. These are consumed hastily, and are only partially masticated. Then the rack is filled with hay or straw, which he can consume at leisure. Later in the evening he is supplied with a quantity of water. Some portion of this passes through the stomach into the intestines, carrying with it a considerable quantity of the unmasticated corn, which passes through the alimentary canal, and is voided in the excrement without having given up any of its nutritive properties. It will be readily seen that under this system of feeding great waste of food is entailed. Next to food, water is of the greatest importance, and a constant supply should be placed within the animal's reach. Where this system is carried out I have never known any unfavourable result to follow. It is in those cases where the water supply is stinted and irregular that excessive quantities, injurious to health, are most likely to be consumed.

To return to the foal. When weaned, it should be confined to a loose box and open yard for the first week, until it has forgotten its dam. It may then be allowed to roam over a pasture during the day, and should, where practicable, be accompanied by others of its own age. Small, well-sheltered enclosures, with high hedges, are best for the purpose. These should contain a shed and an open yard. They should be furnished with a manger, and a constant supply of clean soft water; it is essential that the shed and yard be kept clean and well littered. The pasture should be specially prepared for the foals—a piece of mixed seeds or old pasture, where the stronger-growing grasses are allowed to run to seed. The young animals delight in nibbling off the ripe heads of the grasses. Bare pastures are objectionable, owing to the liability of the young animals to pick up the embryonic germs of objectionable insect life. In order to keep the animals in a healthy state of progressive development, a small daily allowance of artificial food should be given. A mixture of oats, wheat, peas, and a little linseed should be used—the latter not only on account of its nutritive properties, but also for its medicinal and relaxative qualities. All the corn should be ground and mixed with a limited quantity of hay or straw chaff, mixed together and well saturated with boiling water. This should be allowed to remain for not less than twelve hours before being fed.

Castrating the male is an important and, in unskilled hands, a dangerous operation; and in some districts there is a growing disposition to perform it by mechanical means without casting the animal. In many cases the operation is unskilfully and roughly performed; hence the spermatic cord is displaced and injured, rendering another operation necessary before the ligament can be restored. The foal should be trained to lead whilst still sucking, as frequent handling whilst young saves a great amount of trouble at a later period. The colt is generally broken to work when two years old. Every horse should be bitted, and trained to answer to the slightest pressure either when in the plough or cart; if this were more strictly enforced during early life, many serious accidents would be avoided. The colt, when put to work at two years old, requires an extra amount of care in feeding and working. See that the harness fits the animal, and is sound and in good condition, as the skin is as yet tender and is easily galled. A particularly hard or ill-fitting collar frequently lays the foundation of jibbing and other vicious habits. See that the food is sufficient in quantity and of good quality, and do not ignore the fact that the immature animal not only requires a diet sufficient to replace the daily waste of the body, but also to ensure its healthy development. Another most essential point is the feet. "No foot, no horse," is an old adage. The farrier is now being trained to practise a more humane and rational system as regards the foot of the horse. The improved plan is to fit the shoe to the foot, rather than the foot to the shoe. The ancient buttress and draw-knife are less used than formerly; a seaton is formed for the shoe without removing the flaky accumulation from the sole. These act as a buffer, and deaden the concussion caused by sudden contact with obstructions. A very common fault is the use of too many nails, as these weaken the horny walls of the hoof. Another objectionable practice is the too free use of the rasp. This breaks the enamel, so to speak, of the external covering of the hoof, and admits water, which tends to softening and decay. The weight of the shoe in the draught horse is another important point The object of shoeing is the protection of the foot. This can be accomplished by a moderate weight of iron. Beyond this there is an unnecessary waste of power in the case of farm horses, or, in fact, all horses used for draught purposes. Heel calkers are objectionable; high calkers disturb the natural position of the limbs and throw an unequal strain on the muscles of the legs.

There are various pure breeds of draught horses, each of which has a Stud Book for the purpose of registering pedigrees. Of the several breeds, the chief are the Shire, the Clydesdale, and the Suffolk; and during the last fifteen years great progress has been made in the improvement of different breeds. Soil and climate exercise a certain influence, and to a considerable extent fix the type of different races of domesticated animals. This should not be overlooked by the breeder, who in many cases will find it more profitable to improve an indigenous breed, rather than introduce an alien into a locality where it has yet to be proved. Whatever the district and breed, we would strongly urge farmers to carefully register pedigrees. This is more particularly important in the case of mares which in all probability will subsequently be used for breeding purposes. The difference between the value of registered and unregistered mares is very considerable. Except in the case of highly-bred and promising colts, it is safer and, on the whole, more remunerative to castrate the colts and train them on for ordinary commercial purposes. A well-shaped, well-grown gelding is saleable at any age. On moderately-sized farms, where one or two foals are bred yearly, the farm soon becomes overstocked; sales have of necessity to be made at an early age; and thus the breeder is deprived of the best share of the profit. The best geldings must be five years old before they are fit for town or railway work, and arrive at their maximum value. This again brings us back to the use of the sire. Let not the breeder be tempted to use an inferior sire because of a low fee. The produce of a first-rate sire is frequently doubled, even as a foal, whilst the females by a popular sire find a ready market. In these depressed times large landlords can help their tenants immensely by providing a good class of sires. Of this I have ample experience, serving as I do a liberal-minded employer, who keeps for the use of his tenantry different classes of entire horses, at a nominal fee. This has been going on for the last eight years, and the improvement of the horses on the estates is most marked. An important point is to breed from parents free from hereditary disease. It has been too much the custom to breed from old, worn-out mares, or those suffering from disease, the produce of which is always disappointing. As already hinted, breeding from young mares is, other things being equal, invariably attended with the most success. Breed only from registered mares. Many condemn close, or "in and in," breeding; but close breeding can be carried on to a considerable extent without entailing risk of deterioration. When a certain stage has been reached, however, there is greater danger in the selection of an unsuitable outcross than there is in close breeding. Without close breeding you cannot have prepotency, either in the male or female.

The three breeds of horses chiefly used for lurry work in the large towns are the Shire, the Clydesdale, and the Suffolk. The urgency of business requires a more active animal than was formerly the case. A single horse-load varies from two to four tons; hence weight as well as action is required. The class known as "vanners," or those capable of moving at a rapid pace with loads of from one to two tons, is more numerous than that of the other class. Consequently their value is correspondingly reduced. The best matured specimens of the heavy horse readily make from £80 to £100, whilst the best class of vanner may be purchased at from £40 to £50. The scarcity of the best specimens naturally enhances their value. There are no two breeds of heavy horses that "nick" so well as the Shire and the Clydesdale. The former give weight, the latter action. It would be little less than a national calamity to obliterate either the one breed or the other by means of crossing; yet for commercial purposes no breed would be so valuable or pay the breeder better. We strongly urge the owner of one or two good mares to try the experiment.

It is on the grass holdings of the Midland and North Midland Counties that Shire horse-breeding is most successfully carried out. Horse-breeding never pays on large tillage farms, where the mares are working during the busy seasons, as breeding, to be successful, will not admit of working the mares. We believe that climatic influences exercise a more potent effect on the development of animals than the geological formation, which obviously to some extent influences the variety of plant-life with which the soil is clothed. Superficial observers, and those who base their opinions on tradition, assert that animals bred and reared on mountain limestone formations are larger of frame and possessed of more bone than similar animals bred and reared on the millstone grit. We have a practical illustration in the Peak district of Derbyshire. On the north-east we have millstone grit, and on the south-west mountain limestone, the line of demarcation between the two being marked by a tiny rivulet. Here the closest observer can observe no difference in the stock bred and reared on either side. To the tenant farmer whose holding is suitable, the breeding of a few of the best class of heavy horses may be carried on as a remunerative branch of the farmer's business. If well descended and well reared, the horses will be saleable at good prices at any age. Guard against overstocking; and do not lose sight of the fact that horse-breeding, to be pursued successfully, requires the exercise of sound judgment and a considerable amount of close attention. Under these conditions a fair amount of success is certain.

SHIRE HORSES

By Thomas Dykes

SHIRE horses, or the horses of the "Shires," at the present day rank not only as the most powerful draught horses of Great Britain, but of the world. In the Clydesdales and Suffolks, the only British rivals, they have keen competitors for the foreign export market. But when our own street requirements have to be studied,—the movement of loads from dock or railway depôt to warehouse or factory, from mill to bakery, and from brewery to tavern,—the Shire stands out alone. The Scottish and East Anglian breeds have their merits, and for agricultural purposes pure and simple cannot well be surpassed; but when huge drays, laden with merchandise or raw material, have to be taken along our main thoroughfares and suburban highways, we must have animals capable at all times of overcoming the *vis inertiæ* of frequent starts, for the stoppages are as numerous as provoking. We put sail on a ship to make her gather way, and we must put muscular strength between shafts and chains to make the wheels turn round. And so, notwithstanding the improvements in steam road-haulage and the development of electric traction, the ponderous English work-horse, with a history older than that of the English thoroughbred, is in demand in all parts of the world for the purpose of increasing the size, weight, and stamina of the lighter native stocks.

The origin and development of the Shire horse has, since the Stud Book form had its origin in 1877, been treated by many pens. A good deal of theory has been dragged into the work of various writers, who have at no time proved equal to the task of compiling a history with accuracy and literary sequence. It is common to accept in its entirety the modern work-horse as the actual descendant, in a modified form, of the ancient war-horse of the days of chivalry. A certain reservation must be used in regard to this, as the points of excellence required in the latter-day horse of armour were scarcely those which we would admire in the modern dray-horse. What we know from history is that the original British horses, the horses of the Britons who fought against Cæsar, were clever, active, and full of courage. Possibly they remained very much as Cæsar and his legions found them for several centuries after the Roman invasion.

Heavy blood, from the sea-coast of Flanders, was first introduced by King John, who, between 1200 to 1216, imported one hundred Flemish stallions. These stallions no doubt first gave character to our English work-horses. In Normandy the heavy horse of armour was first brought to perfection by crossing the active horses of the north of France with Frisic horses, no horse of the Asiatic breeds being equal to carrying the knights who fought these holy campaigns to battle with the Turks, who despised mail, and relied entirely on the speed of their Arab horses and their own dexterity in handling the javelin. The throwing of the javelin brought the Flemish horses to the front. For several centuries after King John's reign we continued to import into England large Flemish horses; and these, by several enactments, were carefully prevented from being carried across the Border. The Scottish court was, however, closely connected with Northern Europe, and so numerous stud horses and mares were brought into the northern part of the kingdom, possessing fewer of the ancient Frisic characteristics than those which were brought to the south. Here, possibly, originated certain differences in Clydesdale and Shire character, not so readily perceptible to any but those well versed in ancient equine lore. This is a subject which does not fall, however, within the scope of the present chapter. The maintenance of the size of English horses, and the preservation of their best qualities, was a subject which engaged the best minds of the State for at least six centuries, as it does the minds of the richest and most intelligent people in the land to-day.

The horse-breeding farmer of the present day cares little about the Dark Ages period of the history of the Shires, and prefers accounts of triumphs in the modern agricultural show-rings to descriptions of deadly combats in the lists. The Clydesdale farmer loves to speak of the Broomfield Champion, or Thompson's black horse; the Suffolk man, of "Cup-Bearer"; the Hackney breeder, of "Marshland Shales" or Triffit's "Fireaway"; and the Shire horse-breeder, of Dack's "Matchless," and the Packington blind horse—the horse from which the largest pedigree on Shire records is traced.

SHIRE HORSES

The modern Shire horse, as we see him in the Agricultural Hall, owes many of his best features to the development of our great commercial system. Strong and powerful in the field, or on the homestead roads, he was called upon to come out upon the British highways by the carriers and contractors at the commencement of the century, or take up his place on the towing-paths of our canals. Raw goods had to be brought in and manufactured goods sent out. Our ablest engineers gave the same attention to road-making as their successors have done to railroad-making. Then came the railway itself and the iron horse. For a few years it was thought that the great snorting steam steed of George Stephenson would drive the heavy horse into the knacker's yard; but shrewder judges said it would only fetch for the hoofed steed to carry, and would create a greater demand for him for the terminal delivery. Time has proved all these men to be right; and the demand for first-class geldings for heavy street work is practically unlimited.

Without venturing further into stud-book lore, we may say that the most successful sires of the past twenty years—the sires which gave us the horses which are most suitable for work in field or street—have been "Honest Tom," "Lincolnshire Lad," the "Sampson" of the Ellesmere stud, "William the Conqueror," "Harold," "Premier," and Mr. Forshaw's celebrated horse "Bar None." These sires and their stock have greatly helped to mould present tastes, as evidenced in the selection of the famous sire "Vulcan" for double champion honours at the Islington meeting;

or "Starlight," the twice champion mare; and of the three-year-old stallion "Bury Victor Chief," which has drawn honours wherever exhibited, and has never known defeat. Modern breeders go more upon quality than they did previously to the establishment of the English Shire Horse Society. At the earlier shows, animals were found in the honour lists the legs of which were as round as a post and as rough as a terrier. Flat bones have come more into favour, as of better wearing qualities, the set of the pastern has been lengthened, and the hoof made wider and deeper at heel. The rough, "gummy" haired horses which delighted our grandfathers of the fields, who were ignorant of the requirements of heavy street draught, have disappeared, or are only to be found in outlandish corners of England, where the state of agriculture is marked by the weeds and waste in the fields. Since 1870 numerous large breeding studs have been established throughout England, and from these have been sent many sires which have helped to raise the general standard of excellence. The Worseley Hall stud of the Earl of Ellesmere is the oldest, and to-day the pedigree and reputation of the animals, fully considered, ranks as the best. Captain Heaton, the able manager, has no doubt altered his model during the past fifteen years, breeding altogether on "street" lines, and turning out thick-set, blocky horses—legs not too long, but so that the joints are on the level of the waggon axle, when no muscular force is thrown away. Such horses, with flat bones and good feet, do a large amount of work, and last a year or two longer than the long-legged or hairy-legged sort. Mr. Walter Gilbey's stud

SHIRE HORSES

at Elsenham, Essex, has a deservedly high reputation. Originally the animals to be found there were all of the old English waggon type. In Derbyshire the Calwich type holds prominence. Muscle, good easy joints, and a regular walking pace, the distances from hoof-mark to heel-mark being always the same; this ensures steady power in the draught all through the journey.

The Shire horse, as we know him now, weighs in full development a ton and one-eighth. This is the sire at seven years old: he is broad-set, heavy-jointed, and flat in bone. The best Shires are of this order. The worst (there are still many) are round in bone, coarse in hair, thin in hoof, and showing strong evidence of lack of scientific breeding in the forehead, eyes, neck, ears, and general contour. Bones not necessarily flat, but bevelled like a razor, with a fringe standing out therefrom, hocks set for leverage, and knees well cut and carved for powerful forward action, are what we want to-day. The sloped pastern has been called a Scottish feature; it is not necessarily so, for many Shire horses and families of Shires possess this grand feature, sometimes too much treated as a mere showyard qualification. On the streets this point is essential. We like the Shire horse big in frame, powerful in neck, deep in shoulder. From that point we go back to the head, which is Roman in profile, and not so broad between the eyes as the Clydesdale. Coming to the body, the barrel is wider than in the Clydesdale or Suffolk, the back shorter, and the back rib deeper. The quarters and loins ought to be better than we find them. Weak loins, short quarters, angular stifles, lean second thighs and hocks, mechanically or automatically incorrect, are faults to be got rid of. This the presiding authorities in Shire horse matters are doing, and that is the most that can be said in this article. What is great in the Shire is his weight and strength of bone. We want, however, to improve him as regards action and quality.

In regard to breeding sires as a source of agricultural profit, the first requisite of a first-class stud farm is good housing. The American system of having continuous barns with interstices between the boxes has been found to suit well in this country—especially for colts, which seem more settled in company. This system has been adopted to some extent by the Cannock Stud Company, and Mr. Hart, the managing director, has found it to work well. With regard to the roofing of boxes, the Norwegian style of having small gutters running along the edge of the planks, wide enough to let in a little air whilst keeping out rain, has much to recommend it; and the Duke of Hamilton's boxes at his Suffolk stud farm, at Wickham Market, have been covered in this way. The littering of the boxes for young stock (when so much attention is paid in the judging-ring to the importance of every animal having sound, durable hoofs) is a matter for careful consideration. Sawdust, straw, and peat-moss have their advocates; whilst the late Mr. Drew, of Merryton (the Bakewell of work-horse breeding), bedded all the boxes of his young stock with seasand, his contention being that the action of the sand, without causing sand-crack, gradually pared away the horn, this making a naturally-shaped foot.

In regard to the accommodation for mares (if

regularly engaged in farm work, as they ought to be), a large commodious stable, with wide stalls, lofty ceiling, and well-fitted fodder-racks and mangers, will be found more healthy than boxes (which are apt to get stuffy), and in which the isolation is found somewhat unsettling. Some large stud owners who breed for showyard and fashion never allow their mares to be worked at all, being of opinion that haulage somewhat twists them out of general shape and contour. Granted that this is the case, it has to be conceded, on the other hand, that hard work creates muscle, and that the tendency to transmit muscle, in the course of a few generations, becomes hereditary. Moreover, the mare which is kept working will always be found to prove a more certain breeder than one which has merely moved about from show-ring to show-ring, carrying at all times an extra coat of fat in order to captivate the judges. Of course, when so worked, the owner should take careful note of the date of service in his Stud Book, so that no mishap may take place for want of attention. The wax on the ends of the teats will give timely indication of the coming of the foal, and the changing from a dull white or yellow colour to one of transparent lightness should be the signal for the attendant to be at hand. Foals are frequently lost at birth through sheer carelessness; and those who cannot see their way to give their mares a few hours' attention at a critical period, have no right to claim to be considered as horse-breeders at all. The foaling period for cart-horses of nearly all kinds lasts from about the beginning of March till the third week in May. If the foal is dropped in April and the weather is sunny, mother and progeny will be none the worse for being turned out for an hour or two, though great care must be taken to see that the foal does not get chilled. Spring weather is treacherous, and sudden changes in temperature are at all times attended with danger. Corn and hay ought to be given the mare till the grass grows sufficiently long to allow her to pick up her own living. Corn will put heart and style into the mare, and give her coat a gloss; but great caution must be taken not to overfeed, a great many colts and fillies having their powers of propagation impaired in early youth through the mistaken kindness of stud-grooms and stable attendants. The giving of cow's milk is to be generally deprecated, though a practice commonly in vogue amongst those who breed for the prize-ring.

The ninth day after foaling, the mare should be served, and great care should be taken in choosing a suitable sire, regard being had to defects which it is required to qualify or improve, also to the preservation of points which are strong in the mare and her family. Cheap sires should in all cases be avoided, and a proved stock-getter (one that has taken honours at one of the Shire Horse Society's shows at Islington) preferred, even at the charge of an extra guinea or two. When weaned, some crushed oats and a few roots will help on the foals wonderfully; also red clover and rye-grass hay. After they have become yearlings, if the grass is good, extra food may be dropped altogether; and it will be all the better for the growing of bone and hoof if the pasture is on lime-containing soil. The average cost of turning out a good two-year-old fit for work is about £40—made up of service-fee (say three guineas), and the rest keep of mare, less what she contributes to the working bill, and which should be deducted as a set-off. The rations vary as to particular localities and dates of foaling, but the food and grass bill, with service-fee, will generally run to the amount stated. At £40 for a colt rising three, farmers cannot fail to find Shire horse-breeding remunerative, and they have always the chance of breeding some rich nugget like "Bury Victor Chief," "Vulcan," or "Prince William," and making a fortune at a single throw.

LIGHT HORSES: BREEDING AND MANAGEMENT

By Gilbert Murray

The term "light horses" includes a great variety of animals devoted to widely different purposes. My remarks are mainly intended for the information and guidance of tenant farmers, many of whom can breed a colt or two a year without materially interfering with the ordinary operations of the farm. Again, on the low-rented land of the fells the breeding of a popular and useful class of animal may constitute no unimportant part of management, and form another link in the chain of probable income. The macadamised road and the railway have long since annihilated the packhorse, and have well-nigh extinguished that useful animal, the Galloway, from the fells and dales of the Border counties. Some seventy or eighty years ago the sturdy, sure-footed Galloway was the chief means of transport, whether to kirk or market, bridal or funeral.

The breeding of high-class hunters is too speculative an undertaking to be largely entered into by the ordinary farmer. The animal must reach the age of five years before he is of much value to the hunting man; meantime, irrespective of the cost of keep, the risks are considerable. Then there is the trouble and risk of training. Few working farmers have either the time or necessary experience to breed a perfect hunter. It is quite possible for an ordinary farmer, who has a good mare with a dash of blood, to breed a first-flight hunter, but the animal has generally to pass through many hands before he realises a fancy price. Hunters usually pass from the hands of the breeder at a price which, including service-fee and loss of mare's services, barely covers the outlay up to that date. A young farmer who is a good horseman may occasionally have a lucky venture and sell a good-looking, well-mannered four-year-old at a big price, but to the ordinary farmer the game is a risky and speculative one. If he is tempted to try breeding, let it be a sound, well-shaped young mare with bone and action. Be careful to insist on good, well-placed shoulders, with long, sloping pasterns, as without these you cannot have either safe or free action. If your object is to breed a valuable hunter, you must use a thoroughbred sire free from any taint of hereditary disease. Select a sire of mature years to mate with your young mare, and see that his stock come out well; and also study the class of mare with which he "nicks" best. A hunter, to carry 15 or 16 stone, must have bone and substance; but at the same time, whether he is classed as a heavy- or light-weight hunter, he will not command a high price in these break-neck days except he can jump and gallop—in fact, "race" is the more correct word. There is a school which just now advocates the use of cocktail sires; but do not be led away by it. I will recommend the remark of the late Henry Corbett, than whom there was no better judge of light horses: "You are always safe to back Claret on a thoroughbred against Port Wine on a cocktail"; and so the fact remains. By using a cocktail on a half-bred mare you may breed a trapper or light vanner, but never a valuable hunter. We are recommended to select the produce until we get the right sort. A man's life is too short to make much progress on these lines. In this connection, the safest game for the ordinary farmer is to try and breed a useful trapper or vanner, as these are saleable at an early age, and with care and judgment leave a fair margin of profit. In many of the dairy districts, where farmers keep a good mare or two and breed a foal a year, the young one is put to work at two years, generally running the milk to the station twice a day. I recently bought a good lean gelding, barely three years old, which for more than twelve months had daily run the milk to the station, two miles distant. He was probably above the average, as the price (£45) indicates. The demand for light, active vanners and parcel-delivery horses is steadily on the increase, and the price obtained for serviceable animals is remunerative to the breeder.

The fashionable carriage-horse is a whole-coloured animal measuring 15 to 16 hands 2 inches. These are generally the produce of a clean-legged, active, upstanding mare of the Cleveland type, with good crest and forehead, straight on the underline, and long in the quarters. With close attention and a fair amount of judgment, this is a class of horse that pays the farmer-breeder well. The great point is the happy selection of a suitable mare. The young ones are

LIGHT HORSES: BREEDING AND MANAGEMENT

wintered in the straw-yards, and run thinly on the pastures during the summer. When rising three years old, they are bought in their long tails by the London job-masters and livery-stable keepers, who break and match them, and either let them for the season or sell them at high prices. The breeders never break them, with the exception of teaching them to lead in a halter. The large Wold farmers purchase many two-year-olds from their breeders, and put them in condition for selling to the London purchaser. For the small occupier, this is a safe class to breed; and, if not sufficiently good-looking and active for gentlemen's carriages, they make valuable vanners and command the best prices. They must have a large dash of thoroughbred blood in their veins. As a rule, the ordinary farmer grudges the fee of a suitable, sound, thoroughbred sire, hence the progeny is frequently a weedy, worthless animal. Many run away with the erroneous idea that the best way to breed a carriage-horse is by mating an active cart mare with a thoroughbred sire. This seldom fulfils the expectation of the breeder, and the produce is generally soft and sluggish. Even with the best management and most careful selection of the parents, misfits frequently occur.

The term "General Utility" horse implies an animal of mixed lineage, adapted for a variety of purposes. The class of animal in my mind is the active, clean-legged one, 15 to 16 hands 2 inches high, in common use upon the small hill farms of the upland districts of the northern counties of England and the south of Scotland. They are of no special breed, but a cross of different strains acclimatised to the localities in which they are found. Bred on the hill farms, hardily reared, and often exposed to wintry storms, they are usually broken in at two years old. At the age of four or five they pass to town tradesmen, who use them for carts or light vans; whilst the most upstanding and powerful are taken by the tram companies. They are cheaply reared, and pay the tenants of low-rented land well. The mares are seldom put to the stud before the age of three years. There is a good demand for animals of this class at remunerative prices.

The establishment of a Stud Book is not only improving the Hackney, but is giving it more prominence and greater popularity; and the best animals are in good demand at high prices. There is a growing demand for animals of greater size—and 13 to 15 hands 2 inches are most saleable. Originally there were two distinct varieties of the Hackney: the Yorkshire type, with good forehead, and sloping shoulders, well laid into the back; long, sloping pasterns; hard, clean legs; flat bone; and well-developed muscles. This was the true Hackney type—better adapted for the saddle than for harness. The Norfolk is shorter in the neck and thicker in the shoulder, with pasterns shorter and more upright; and hence not so pleasant for saddle-work. The best Hackneys of the present day are a cross between the two strains, the difference in the shape and appearance being due to climatic influences rather than to race. The breeding of Hackneys may be taken up by the ordinary farmer with profit. Select a shapely mare to start with; use a well-selected pedigree sire: by continuing this selection on the produce the result soon becomes apparent. Every cross enhances the value of the next generation; and by the exercise of ordinary judgment there is little risk attending the breeding of Hackneys. Promising young horses are readily saleable at any age. The service-fee is the greatest item, and parsimony on this point is not economy. The name and reputation of a sire makes a considerable difference in the value of the stock. There is much less risk attached to the breeding of Hackneys than is the case with other varieties. They need not be expensively kept, and at the age of two years are capable of earning their keep and of fulfilling a variety of useful purposes. Under 14 hands, they are ponies; from 14 to 14 hands 3 inches, they are cover hacks; 15 hands and over, they are hackneys, and are suited either for saddle or harness. With well-balanced points and free action, they make a ready sale at good prices. Some prefer high-knee action; but for safety and durability we prefer the straight action from the shoulder, and whilst in motion the hind-legs should come well under the body. The motion of a good Hackney should exactly resemble that of the hare, the true character of which is exemplified by its footmarks when traced on a slight covering of snow.

The standard height of the Polo pony is 14 hands. The best specimens have a strong dash of Barb or thoroughbred blood in their veins. They should be rather long in the neck; shoulders fine, and sloping into the back; fine in the wither gaskins; well let down; hocks flat and strong, slightly inclining inwards. High action is objectionable, though, when required, they should be capable of going at racing speed. Some of the most valuable animals in the Kingdom have been bred and trained by the Earl of Harrington; they are the produce of well-selected pony mares, by a Barb sire. This gives them strength, endurance, and speed. Without wearing, high action, these ponies could be profitably bred by the occupier of mountain lands in all parts of the kingdom where the services of a suitable sire can be obtained. To be successful, you must either have recourse to Eastern blood or to an undersized thoroughbred, in which case there is the danger of many misfits. When matches were on in the country I have frequently seen from twenty-five to thirty Polo ponies at Elvaston, belonging to Lord Harrington and others, worth from 100 to 250 guineas each; I need hardly say that these were "made" ponies, and the pick of the basket. They can only be trained by an experienced polo player; but the breeder may handle and form their mouths at from two to three years old. At this age good specimens will make from £30 to £50 each; and no great expense need be entailed in their breeding: all that is necessary is an ordinary hill pasture in summer, removal to lower ground in winter, a shed to which they can resort in severe weather, with a moderate allowance of oats and hay, or, what is still better, a daily allowance of chopped hay and mixed meals. These must be placed in the manger of the shed, in order to prevent waste from the weather. Here again you have an animal which, if not up to the standard of a first-class Polo pony, is yet valuable for many purposes: no animal is so saleable as a good fourteen-hander.

LIGHT HORSES: BREEDING AND MANAGEMENT

On most farms the maid-of-all-work is the pony. Some one wants to catch a train, some light packages have arrived at the station, or a visit to the village or the market is contemplated, when the word passes round, "take the pony." Or it may be the ladies of the family wish to make a call, and if too far to walk comfortably, take the pony-carriage: the animal is ready for any and every emergency. The term pony includes all heights from 12 to 14 hands; its native habitat is in the mountain regions of Wales and the far north of Scotland, whilst numbers are bred in the New Forest in Hampshire. Large droves are imported to the lowlands both of England and Scotland at the age of about two years, and these are bought by the farmers at a low price. On the richer soils they increase in size; many grow into extremely serviceable animals, and the inferior grades are used chiefly in mines. The use of better sires has done and is doing much to improve the race. We cannot recommend the breeding of animals of this class on enclosed and cultivated lands; nevertheless, when purchased at the age of one or two years, they will pay well for a year or two's keep. A cross with an Eastern sire has in some cases produced good results; but the introduction of a remote out-cross requires caution and discrimination. At certain seasons there is a good demand for strong sure-footed animals of this class, for the use of sportsmen in hilly countries. Those who follow sedentary occupations for nine or ten months of the year are not in condition to take long journeys on foot; and many who are infirm enjoy the sport which can be had from the back of a strong, quiet pony. The lowland farmer can grow and train the animal, and in this way realise a profit, instead of embarking in his breeding.

Villa residents on the outskirts of large and densely-populated districts use ponies for their children for the purpose of obtaining fresh air—a small pony, on which a young hopeful is swung in a pannier on each side. These animals are generally of Lilliputian stature and throughly domesticated. Many are of Norwegian origin; they are sluggish in temperament, and their general appearance discloses their plebeian origin. Except they be of superior class, their breeding would not pay the English farmer. In parts of the Highlands, and in some of the Islands of Scotland, two-year-olds can be bought for a few pounds; and even then the cost of keep and the trouble of breaking leaves but a meagre margin for profit.

SHEEP

By Professor Sheldon

ANCIENT documents contain such sparse and fragmentary references to the sheep of the period that we are unable to decide, with anything approaching certainty, which of our breeds of to-day can claim the greatest antiquity. The probability is, however, that the moor and mountain breeds—the heavy-horned ones—represent most closely the types of sheep indigenous to prehistoric Britain. The Dorset, the Exmoor, the Black-faced—more especially the last-named, perhaps—have in all likelihood made the least change from the ancient pattern. I conjecture this because the "Big-horns"—the wild sheep of the Rocky Mountains of British Columbia, whose origin can only be imagined, and that with vagueness—have horns like these, and resemble them more or less in general outline. Mountain sheep, indeed, are more likely than any others to have survived the dangers of a feral state; but this, again, is mere conjecture. It is at all events inferentially true that different breeds did live through the period when wolves were not uncommon in this country, for it is extremely improbable that the Romans introduced them. No; the sheep of England are indigenous, and the varieties are owing to habitat and natural selection.

It is quite possible, however, that the Down breeds in the southern counties may possess an antiquity of blood equal to that of the horned sheep of the Fells; and the Sussex Downs, at all events, are admitted to be as pure-bred as any we possess. This breed is now universally known as the Southdown, because it has spread a good deal over the southern counties; but it is just as well to indicate the county to which it more especially belongs. The Down breeds—those of Sussex and Hampshire—and the breeds of the moors and mountains are understood to be indigenous to their respective habitats. These downs and moors were open spaces of the country, and the primeval sheep, not being animals of the forest, took naturally to them. The original variations in the breeds of sheep were the result of natural selection, of soil, and of climate; but the modern sub-varieties have been evolved by artificial crossing and selection.

The Southdowns are a breed of great antiquity, and their distinguishing characteristics are fixed; hence their prepotency when crossed with other breeds. They have been long celebrated for the fineness and closeness of their wool and for the close-grained character and superior flavour of their mutton. In these respects, conjointly, they are not equalled by any other sheep in this country, or indeed elsewhere. They are, however, too small to be as profitable as some of the sub-varieties; but on the thin soils and sparse herbage of the chalk downs they find sustenance where most other breeds would starve. Being light and active, and full of restless energy, they are able to roam over large areas in search of food where necessary. At the same time they readily adapt themselves to better land, to small enclosures, to cultivated crops, and are found to develop in size under these conditions. In this way, flocks of great excellence have been and are being produced in various parts of the country. Their influence in the improvement of various other breeds has been remarkable, and they are still used for this purpose to some extent.

The Hampshire Down, in its present form, is quite a modern breed, limited in point of time to the present century. The Hampshire Downs of to-day are the product, originally, of crosses between the Wiltshire Horned-sheep, the Berkshire Knot, and the Southdown, and were subsequently improved by careful selection and judicious management. Their character is now sufficiently fixed to be regarded as permanent and unvarying. Their quality of wool and mutton is owing chiefly to the admixture of Southdown blood; and their greater size to the other elements in the "cross." Better, perhaps, than any other breed, they adapt themselves to close and continuous feeding on roots and green crops generally. There are no better nurses than the ewes of this breed, and their lambs attain considerable size at an early age. Their wool is close and fine, though not equal to that of the Southdown, and their faces are black and Roman-nosed. On account of their fecundity and superiority as nurses, the draft ewes are in good demand for crossing with Lincoln and Cotswold rams for the production of uncommonly vigorous cross-bred lambs, which are fattened early for the butcher. From the sires these lambs derive increased size, and from the dams superior qualities of flesh. In regard to no other breed of sheep are breeding and management carried on more scientifically and intensively, if indeed as much so, or on so large a scale. The annual sales and lettings of rams on many of the most famous breeders' farms are institutions of a large and important nature, attracting crowds of farmers from far and near, and the prices realised frequently run high.

In Shropshire Downs we have a valuable breed of comparatively modern derivation, the most important characteristics of which are inherited from Southdown crosses with the indigenous horned sheep of the Longmynd district. The wonderful success in the breeding of improved Leicester sheep which was attained by Bakewell of Dishley, in the latter part of the eighteenth century, gave an impetus that was greatly needed to sheep-breeding generally. To this famous man, indeed, is due the credit of setting to the farmers of England an example in the breeding of improved sheep, cattle, and horses, which has been, and is, more potent for good than that of any other man. To this example, indeed, we may fairly attribute the evolution of such sub-varieties of surpassing excellence as the Hampshires, Oxfords, and Shropshires; and he may be said to have himself virtually created the Border Leicesters. Of the three Down breeds, the last-named has been more widely disseminated than the other two, and in that sense it is the most important. Shropshires, indeed, have the faculty of thriving well on a greater range of soils, and they are found in many counties and countries where the others are scarcely known. They are hardy and docile, excellent breeders, and their mutton and wool together are superior in quality, if not in quantity, to those of any other home breed, save Hampshires and Southdowns. Where sufficient attention is devoted to their breeding and management, the Shropshires are bold, handsome, symmetrical sheep. Their popularity is increasing, and will continue to increase, alike for use at home and for export to many foreign countries.

The Oxford Down is the largest and, in some respects, the noblest-looking of the Down breeds, of which it is the latest—that is, the most modern— development. It is understood to have been formed chiefly by crosses between the Hampshire Down and the grey-faced Cotswold, with a few dashes of Southdown blood for quality. As a recognised breed it is not yet half a century old, but in the hands of such men as Mr. John Treadwell it has attained a very considerable measure of fame and popularity. These sheep are considered, under suitable conditions, to be as profitable as any other breed, on account of size, weight of wool, aptitude to fatten, and vigorous and healthy constitution. They are, however, as may be expected from the admixture of Cotswold blood, coarser than the other Down breeds, both in wool and mutton. They are more cosmopolitan in adaptability than Hampshires, but less so than Shropshires, and it may be doubted if they are suited to so large a variety of soils as the last-named. All the same, however, they form one of a quartette of Down breeds whose popularity is destined to go on increasing, and they are essentially a breed which merits all the pushing and good management that can be devoted to them. How far Down sheep will emulate Shorthorn cattle in displacing other breeds remains to be seen; but it would seem that, in the race for the survival of the fittest, they are eminently suited to take the place, to some extent, of the Cotswolds and Lincolns.

The sheep of the Welsh mountains, whose mutton has for a long period been popular with epicures, are for the most part white-faced, though some are grey, others mottled, and yet others brown. They are smaller than any other breed in the Kingdom, and their clip of wool is only about 2 lb. This hair-like wool is the result of climate and exposure; and even the Southdown wool would doubtless become hairy, if the sheep could survive whilst the experiment was being made. This change in the character of wool, be it understood, would not occur in one generation, or two, or half a dozen perhaps; but whilst fine wool is a natural attribute of sheep bred for centuries in a genial climate, coarser and hairy wool is produced in cold districts and on the mountains. The wool even of the Leicester (limestone) sheep of the Peak of Derbyshire reveals in some places a tendency to revert to hair. More than any other animal, as it would seem, the sheep is susceptible to change of character, in obedience to different soils and climates.

The black-faced, or mottle-faced, and horned mountain sheep are found on the wild heather-clad mountains of Scotland, and the northern counties of England, where no other breed in England would gain a living, not even the Welsh mountain sheep, in all probability. They are found on the wet grouse moors even of Derbyshire and Staffordshire, and grimy little sweeps they are. This coating of grime comes off the heather through which they push their way, and in some districts is the result of a smoky atmosphere. The sheep of any elevated district, whatever their breed, usually become more or less grimy in the winter, even where no heather is. It must be owing to the fungus or to the smoke which gathers on the decaying grass bents—to one or both of these,—and, in part, to the severity of the winter climate. The sheep of Ireland do not blacken like ours; but then the Irish climate is mellowed by the Gulf Stream, the smoke of peat is not so defiling as that of coal, and there is no great amount of it in the atmosphere of the Emerald Isle. I have mentioned this question of grime on wool, as it is one of some interest.

Mountain sheep possess a good deal of antiquity, and they have an old-world look about them. They are probably the oldest type of sheep in Britain, and less altered than any of the other breeds from what they were originally. They differ more or less in one district from what they are in another; and in the south, generally, from what they are in the north. In some localities they have been more carefully bred, and are therefore of an improved type; in others, no care whatever has been taken, and the sheep are unimproved. But however this may be, they everywhere fill a position of usefulness, living where rabbits hardly would live, and nursing their lambs with a self-denying fidelity unequalled by other sheep. They are as agile as goats, and therefore unsuitable for enclosed and cultivated land. The quality and flavour of their mutton are considered superior to those of the white-faced breeds, but the sheep are small. The wool is not of much account, because of its half-hairy character.

Leicesters are the most famous of the white-faced breeds, not because they are more excellent than Cotswolds or Lincolns, but for the simple reason that Robert Bakewell, a century ago, achieved such phenomenal success in breeding and improving them. It will remain a source of regret that he did not try his hand at creating a new breed concurrently with improving an old one, and that he had not the Shorthorn to deal with instead of the Longhorn. It is probable that, had his attention been directed to the formation of a new breed or sub-variety, he would have accomplished it with marked ability; for it is said by some that he really created a new sub-breed out of Leicester blood, and that it is represented by the Border Leicesters of to-day. Were he here now we may surmise that his incomparable skill would be devoted, not to Leicesters and Longhorns, but to the Shropshire sheep and Shorthorn cattle, as the two breeds of our food-producing animals which have proved to be capable of the greatest achievements and the widest dissemination. In spite of the abnormal fame with which, for the time being, he endowed Longhorn cattle, the breed is dying out. This is not the case with Leicester sheep, and probably never will be, although they have fallen a good distance behind the Down breeds in popularity with those who eat mutton, and it is hardly possible that in the future their standing will be maintained. Like the Southdowns, however, they have had much to do with the improvement of other breeds, and their blood exists in all, or nearly all, the white-faced breeds of England.

It cannot be denied that Leicesters are a practical and hardy breed, suitable for a variety of soils and districts, and for different methods of management. On all sound soils they are at home, preferring grass land to arable, but answering fairly well to folding on root and green crops, although not so well as some of the Down breeds, or even Cotswolds and Lincolns. They cut a fleece inferior in quality to that of some of the other white-faces; and their mutton, inferior to that of the Down breeds, is as decidedly superior to that, for instance, of the Cotswolds. The faults of Leicester mutton lie in its surplusage of fat and its deficiency in flavour as compared with dark-faced mutton.

The ewes are not so prolific as the Hampshire, nor as good nurses; they are just average sheep in these respects, and nothing more. They thrive, however, in rigorous climates, in which it would be cruelty to place any of the Down breeds, save perhaps Shropshires. They are the prevailing breed on the upland soils of North Derbyshire, which are on the carboniferous limestone, and are commonly enough called "Limestones" there. These soils are sound, but for the most part exposed to all the winds that blow, and are not as a rule heavy. The winters are commonly severe, and the springs sometimes late in coming, but the sheep weather the storms remarkably well. As a rule, they allow themselves to be kept within bounds by the stonewall fences; but if once they begin roaming, few fences prove an insurmountable obstacle to them.

The Border Leicesters, so named because they inhabit the Border counties of England and Scotland, are understood to be descended from Bakewell's improved Leicesters, but crossed in some cases more or less with Cheviot blood. They are said to have been first introduced by Messrs. Culley, a century and a quarter ago, and after them by various others. Persistent attention has been paid to their breeding ever since, and now they are considered, as a breed, to be superior to the prototype of the Midland Counties. The purebred Border Leicesters are considered the most valuable, and the tups are in great demand for crossing with Cheviots and Black-faced mountain sheep, for the production of mutton. The first and second crosses produce vigorous sheep, superior to either pure breed as butcher's sheep.

Cotswolds. Deriving its name from the Cotswold Hills of Gloucestershire, the Cotswold is said to be one of the oldest breeds of sheep in England. The wool of these old-time sheep, inferior to what it is to-day, no doubt, was held in great repute. When Henry VI. was on the throne of England, the King of Portugal applied to him "for permission to export sixty sacks of Cotteswold wool, in order that he might manufacture certain cloths of gold at Florence for his private use." Cloth-making was carried on at Gloucester in the days of the Saxon Heptarchy, most probably of Cotswold wool; and these sheep were chiefly valuable, even in those days, for the length and quality of their staple. The breed of to-day is of course greatly improved by careful management, judicious selection, and close domestication. A cross of Leicester blood is supposed to have been introduced in Bakewell's days, with the object of fining-down some of the coarseness of the Wold sheep. The Cotswolds of to-day are fine, handsome sheep, remarkable for symmetry, early maturity, and size. The mutton, however, is as rude in flavour as it is coarse in texture, and the wool is not held in the regard it once was. To those who have been used to Down mutton, especially to Southdown, the Cotswold compares with striking disadvantage. All the same, however, it has its admirers, and finds the needful customers. In many cases the sheep are closely-folded on arable crops; and in wet weather, inches deep in the sticky oolitic soil of the Cotswold hills, they often present a woebegone and melancholy picture of hopeless misery. But there is more in the look than the reality; for as a rule they are well fed, and mud is not so objectionable to them as we might think.

The Long-wooled Lincoln is an old breed, greatly improved in modern times. It is even capable of greater size, though not of handsomer looks, than the Cotswold. The Cotswold top-knot adds much to the appearance of the sheep; and the bald heads of the Lincolns show, by way of contrast, how much the addition amounts to. The old Lincoln sheep were long-legged, bony, ungainly, and altogether undesirable animals. Now they are shapely, massive, well balanced, filling the eye, if not pleasing it, more than others do. The improvement effected is owing to careful selection of sires and " a judicious admixture of Leicester blood." In this way was created an entirely new type of sheep, which retained the pre-eminent wool-bearing property of the old breed, combined with a marked improvement in form, and aptitude to accumulate flesh. In this improved form the Lincolns are to all intents and purposes a distinct breed, but were not recognised as such by the Royal Agricultural Society until less than a quarter of a century ago. Seen at the Royal or Smithfield Shows, they are striking chiefly on account of size, but also because of their curly wool and massive symmetry of form. They are prolific breeders, hardy, and vigorous, well adapted for the bleak wolds of their native county, and habituated to simple methods of management.

The Devon Long-wool is a breed similar in appearance to the Lincoln, but smaller. The best flocks have been derived from a cross between the Leicester and the Bampton—an old Devonshire stock. The Leicester type prevails, and indeed the old Devons were much after the style of Leicesters. But the breed is local, and highly popular in the county, some of the breeders considering them better than any other variety—at all events for Devonshire.

The Cheviot is essentially a breed of the hills, active and hardy to a degree not equalled except by the sheep of the mountains. For a long time they were bred only on the Cheviot hills, but they are now to be found in many districts. They are said to have been much improved by a cross on the old Lincoln sheep a hundred years ago. For the cold—not the coldest—sheep-runs of the north, they are found to be exactly suitable. The coldest ground is fit only for the black- and mottled-faced mountain sheep, which thrive where any other breeds would starve and perish. The mutton of the Cheviots is considered almost equal to that of mountain sheep, but inferior to that of the Down breeds. The ewes are prolific, and good nurses. Draft ewes are in constant demand for a final crop of lambs by rams of larger breeds, when, with their lambs, they are fed off for the butcher on farms where "flying flocks" are used,—that is, flocks which are not kept on for more than one season, but are replaced by others, which in turn are similarly disposed of.

The Dorset Horn, although heavily horned, is not a mountain breed, but it is difficult to believe it did not come originally from the mountains,—or it would be difficult were it not for the fact that it is white-faced. The breed is seldom to be found in its original state, for most farmers have crossed it more or less with breeds in adjoining counties, and in some cases with the Southdown and Leicester.

The value of the Dorset Horns lies in their fecundity: the ewes are prepared to breed at any time of the year, even twice a year if required to do so. They are consequently important, as the breed which produces the earliest lambs for the butcher. They are not at all delicate or fastidious sheep, and are found to be suitable for rough, sour land, on which any other southern breed would deteriorate. The horns of some of the older rams are handsomer in curve, and larger, than those of any other British breed, the Scotch mountain sheep not excepted, and are much in demand as ornaments for entrance-halls, etc.

It is not for mutton alone, or for wool alone, that we can afford to keep sheep on medium and good land; we want both, each of them good of its kind, and enough of both in reason. On the mountains only mountain sheep can be kept, and they yield a profit where else there would be nothing but loss. But on the uplands and lowlands, on the downs and in the valleys, the sheep kept should all be good of their kind —as good as they can conveniently be bred—good for mutton and for wool. It is hopeless to expect to breed White-faced mutton of quality equal to that of the Down breeds, or White-faced wool either; but a great deal of White-faced mutton and wool are far below what they might be in quality. This is true, too, of the Down breeds; and then it becomes a question of feeding and management, equality of land, shelter, altitude, and so on. Every sheep farmer knows which are the best breeds, in respect of mutton and wool, or if not, the markets will soon tell him.

The public taste seems of late years to have been running more on beef than mutton as an article of food. In any case, we have two and a quarter millions more cattle, and two millions fewer sheep, than we had in 1868, and our imports of beef are greater than those of mutton. In other words, more beef is raised in this country, and less mutton, than was formerly the case; and, as our imports of beef exceed those of mutton, more beef is consumed, relatively as well as actually. All this may possibly change in a few years' time; no one can say it will not. The working classes consume more flesh meat by far than they did in the middle of the century, and they have a settled impression that beef is stronger food than mutton—better worth a given price per pound. This is an intuition which is probably sound and true, and, so long as it lasts, the masses of the people will run more on beef than on mutton.

It must be admitted, I think, that the competition of the Down breeds is running the White-faces closely in these latter years. To a great extent it is a question of quality of mutton, which is higher in the dark- than the light-faced breeds. The Downs are quietly displacing the White-faces in localities suitable to the former breeds. But they can never wholly displace them everywhere. It is to be hoped, indeed, that we shall preserve all our breeds, and still produce others of the character of the Hampshires, the Oxfords, and the Border Leicesters. A good deal of cross-breeding is done, but not often with the object of producing a new and distinct variety. Crossing is done chiefly with the object of producing lamb and mutton for the butcher, and the established breeds are necessary to this. In order to improve these, the practice of breeding only from good specimens must be extended. All lambs of an inferior character in any given breed should be fed off sooner or later, and not be put to breeding at all. To a great extent this has long been the case with the males, and it should be extended to the females too. Too many inferior rams are put to service, and yet there are plenty of good ones to be had for the buying. A couple of sovereigns are, as a rule, saved to bad purpose in buying an inferior ram. If only good rams were at any time used, the ewes would take care of themselves, and a general improvement would be its sequel. In many districts the sheep are about as good as they can be, while in others they are not. They ought to be as good as they reasonably can be, and this is soonest accomplished by using superior rams instead of middling ones.

In the management of sheep it is a mistake to have them too thick on the ground. Breeding flocks want room; and they must be fed accordingly. It is surprising to see the number of sheep which some of the arable farms in the south carry. Hampshires will bear crowding like Chinamen, but they want good feeding and a frequent change of food. It is impracticable to farm like this—intensively—on grass land, for there the sheep must have room, and the grass must have time to sweeten. But even on grass land sheep farming will pay for generous feeding, or else it will not pay at all.

PIGS

By Professor Malden

The British breeds of pigs are distinctly superior to those of other countries, although this position could not be claimed for them until within comparatively recent years. It does not appear that there is any special reason for their superiority, beyond the fact that their improvement has been taken in hand by men specially fitted to deal with them; and these men have acted on good lines. It is of course impossible that the creating of the breeds should have been carried out without mistakes; fancy, or the special preference of particular points, at times leads a man to place too high a value on a feature which in reality is not so valuable as he deems it; but the good sense of the majority has been strong enough to prevent permanent injury being done. It is easy to look at some highly-bred specimens of most breeds and fix on a point which, in the opinion of good judges, constitutes a deficiency; at the same time, by judicious mating, that weakness is not so deep-seated that it cannot be eradicated in two or three generations.

The three Yorkshire breeds—the Large White, the Middle White, and the Small White—present a very good illustration of the manner in which public opinion

—especially as regulated by the market demand, commanded by the popular taste—has influenced both "fancy" and more practical breeders. Breeders saw that they must adopt the lines of those who had improved the breeds of cattle and sheep, and aim at early maturity, even though it was accompanied by a diminution in size. To a great extent early maturity is synonymous with quality, though this may be, to some extent, overdone. The original Yorkshire was a pig which, when matured, weighed a thousand pounds, certain specimens turning the scale some hundreds of pounds above that weight. But this was only obtained after a long, slow, and costly feeding, which left very little profit in the transaction, and then resulted in a coarse carcass of not too agreeable meat.

The first aim was to bring this pig to a more rapid maturity; and this, with improvements from time to time, resulted in the modern Large White breed, which, though not so large, has proved far more valuable. In size it is still fully large for the modern taste in pork; but the undesirable coarseness has been done away with, and it has become highly popular in other countries, as well as in England. It must always be borne in mind that quality is not entirely regulated by size. A small animal may possess coarseness, though there is a general tendency for quality to accompany smallness. Seeing that there was a popular feeling in favour of quality, it is not surprising that the breed-maker should aim at a type in which quality should be supreme. The breeders for quality, therefore, set before them a standard in which size was to a considerable extent ignored. Perhaps there is no animal which has been developed to such perfection as the Small White, though it is not necessarily the most profitable. It is difficult to conceive an animal which, from a porcine point of view, could be more perfect in outline, lighter in offal, or more rapid in attaining maturity; it is hardly beside the mark to say that it is in a state of maturity from birth. There is, however, another view of the matter: the fining-down has resulted in an animal which lays on more fat than is desirable; consequently it is charged with losing too much weight in cooking. When very highly-bred its constitution is not too strong, and this is evidenced more particularly in its early life. Between the Large White (although it now possesses much in common with the Middle White) and the Small White there is obviously a wide gap—they may be looked upon as two extremes; it is not strange, therefore, that those who saw the advantages and disadvantages of the two types should conceive that there might be a happy mean: one which should combine early maturity with size, and quality with robustness. With means at hand this was not difficult; and now the popular Middle White must be acknowledged to be a pig suited to the taste of the consumer, and profitable to the producer. These developments have not been obtained without heart-burnings; but the outspokenness of those whose fancies lay in different directions caused every point and feature to be thoroughly thought out, and at the present time we benefit from the care bestowed upon the animals.

Although the Middle White is the standard at which most breeders may be inclined to aim, it is not to be taken that the two other white breeds have not their place, or are not worthy of a place, on the farm as well as in the show-pen. There are features in them which are well worth preserving, and without which the breeder would be worse off. Generations of close breeding and excessive feeding of the Middle Whites, unless great care is exercised, will tend to develop in them some of the objectionable features possessed by the small breed; and it will be found desirable to bring in a dash of the large breed, to give it additional robustness, increased size, and probably greater fecundity. As a matter of fact, in all breeds where too close breeding has been observed, it is found necessary to introduce outside blood, animals of very widely different breeds being drawn in to rectify the weaknesses which have been developed. This is imputing no dishonesty to breeders; they find that in aiming to secure one improvement they have sacrificed something which, for the good of all, must be regained. Those which are not true to type at the time of birth are done away with, and generally in the course of three or four generations there are no " throw-backs," the herd is practically pure, and decidedly more valuable. If the improvement desired can be achieved by mating with a breed which possesses many characteristics in common, so much the better. In-breeding always develops weaknesses when it is carried beyond a certain point, and these must be guarded against. The smallness of the Small White breed is not the result of in-breeding so much as of selection, and the concentrated excellence of some of its characteristics are of great value in mating with less well-bred animals which are excessively inclined to coarseness. It must not be forgotten, however, that popular fancy tends to make the big breeds approach the little ones, and the small ones the bigger.

Among the coloured breeds the Berkshire takes very much the same position as the Middle White among the White breeds. It strikes the happy mean, though there is no breed which can be placed exactly in the position of the Large White. The Tamworth more nearly approaches this, but it is only distantly. The Small White breed is represented by the Essex, Suffolk, and Dorset black strains. The combination of size and quality place the Berkshire in the premier place among the coloured breeds, for it also matures rapidly. The Tamworth and the three small black breeds are very widely divided; the latter come to a rapid maturity with a considerable proportion of fat, but a minimum quantity of offal. The Tamworth matures slowly, is coarser in the offal, but provides meat of excellent quality, there being a nice admixture of lean among the fat. The fat, too, is not gross. It must be remembered that the fat of different breeds varies in quality. Some breeds, particularly those with an undue mixture of Neapolitan and Chinese blood, develop fat of a more greasy and less palatable nature than those which have only a fair amount of these in them, but which have been perfected more by selection. Of course, the coarsest strains are not included among the latter.

A farmer or pig-keeper has to decide upon the class of pig he requires very much by the local demand, and the food he has at his disposal. There are some districts in which few pigs are grown to be converted into bacon, while in others bacon is the chief object of the feeder. Again, colour largely influences the sale. In one district white pigs are unpopular, while a few miles distant black ones are not in demand. Generally, in the south black is preferred, while white is more popular in the northern districts. Yet where good herds of either have been maintained for a long time, and their good points have been recognised, a local preference has developed which has overruled the general taste. This is due to the fact that all breeds possess good features, and that soil, and to some extent climate, have little to do with their thriving. Cattle generally do much better on particular soils, and under special climatic influences. Herefords do best on the red sandstone and throughout the Midlands than in districts outside them. Jerseys prefer the milder climate of the south to the cold of the north. Hill sheep do best on their native heath, while long-woolled breeds come to greater perfection on rich lowland pastures. Breeds of pigs are but slightly affected by their surroundings, consequently the pig-keeper has little to consider on these grounds, and he can choose his breed in accordance with the market demands.

A large quantity of salted pork, ham, and bacon is imported, so that, although in some districts of England there is a large quantity of bacon made, the fresh pork trade is of very great importance. The altered condition of the artisan and the labourer has placed it more in their hands to select the class of meat they desire, and this has been very much felt in the pork trade. The coarse joints they used to patronise have had to give way to the more nutritious and more palatable small joints, in which there is more lean, and the fat does not shrink so much in cooking. Even in bacon the taste has changed, the once admired heavy sides of coarse fat, such as may be remembered by those who lived in those districts where the original great hog was found, no longer continuing to be the pride of the owner. If those farmers who used to find it profitable to give this class of bacon to the men who boarded on the farm were to place it before them now, they would find it objected to very strongly. In this way it has been forced upon feeders to keep before them the necessity of keeping an eye to quality. The market quotations make this more plain, and it is the practice of large curing firms to offer prices graded systematically by weight of the animals, starting as low as from seven to eight score, and taking off about 6d. for each score until a uniform grossness is reached.

In selecting the breed he will affect, the farmer therefore aims at one which will attain maturity at eight score if required, though he does not mean to tie himself down to that if circumstances show that it will be more profitable to grow them to greater size, should the demand call for it; but, as a rule, he wants to have pig which, after that weight is reached, may be made up for market in a short time. It is for this reason that Tamworth pigs do not answer the purpose of the farmer so well as do some of the more rapidly maturing breeds. Instead of topping-up rapidly when put on to better food, it has a tendency to increase in size in place of becoming fit for the butcher. It is not so much a general-purpose pig; yet it is capable of producing bacon of the highest quality, with the meat plentifully streaked with lean, and the fat of most agreeable flavour. Were the sole aim the choicest quality, the Tamworth would in all probability be the most popular. Those who wish to produce a small quantity of bacon for their own consumption, and are not strictly regulated by the cost of its production, could not do better than keep Tamworths.

The Berkshire and the Middle White stand out as the best general-purpose pigs, as they can be brought out at almost any period of their life. They are well adapted for making into pork and bacon. Where not too finely bred, the sides are long and deep; but in aiming at quality there is no doubt that some breeders have gone a little over the mark, and have developed an animal too fat for making bacon of the finest quality. The breed commonly know as the Bedford is very closely allied to the Middle White, and is not the descendant of the native breed; it is traceable to a few breeders in Bedfordshire and Huntingdon who have held herds of the Middle and Large breeds, from which it has been developed. The smaller breeds are valuable as pork; for, if fat, they are delicately flavoured and nutritious, capable of being cut into small joints suitable for ordinary domestic use. They also produce small hams, which find favour in houses where larger ones are not so profitable. The Large White breed is essentially a bacon pig, especially since it has been freed from the charge of coarseness. Its deep fleshy sides are of great length, affording a large surface of valuable food in which there is little offal, thus rendering the offal much less in reality than might be expected at first sight. The large hams find much favour in many districts, and altogether it is a profitable pig to keep.

Speaking generally, the larger breeds are more adapted for those places where there is a great quantity of offal food of moderarely high-feeding properties. On such food it is more profitable to grow big animals; they turn to useful account offal food and build up a frame at little cost, while as they approach maturity they can, with the addition of a little extra food, be made fit for the butcher. If the price per pound is not quite the highest, it is counterbalanced by the bulk and the relative cost of production. Such animals are well suited to run in farmyards, where food would otherwise be wasted, or for feeding on clover and other green crops. They are therefore held in high opinion by farmers. Small pigs are not so profitable in the yard, and are usually not so prolific. They require careful feeding when young, as over-feeding is liable to bring on apoplexy; their housing also must be warm, as they are to a certain extent liable to chills.

The farmer most often requires pure-bred animals to mate with those less well bred, to correct deficiencies which exist among his own stock. Unthriving pigs want mating with those breeds which come to quick maturity; while if they are stunted in build, he requires those with longer frames.

As we are not dealing with the pig as a showyard animal so much as from a farmer's point of view, without going into too detailed a description of the points of the breeds common in England, it may be advisable to mention a few of the most important, and give reasons why value is attached to them. The Large White pig should have a head which is wide, deep, and of fair length: a full head denotes that the animal is of a quiet disposition, not inclined to exercise itself unduly, while the cheeks are considered a delicacy. The nose should not be too much of a pug, because, except in those instances where the animal is fattened from birth, this breed of pig is suitable for searching for a portion of its food, until it is finally put into the sty to be got ready for the butcher. The long frame can be built up on a comparatively unconcentrated diet. The pig is suitable for running in the yard, where it picks up much that would otherwise be wasted. It is particularly suited to run on clover, or to be fed with clover, vetches, or roots which are brought to it. It is little likely to be injured in the yard by other animals, as it is strong and active enough to take care of itself. For a long period of its life it can thus be maintained at little cost; and when its frame is made up it quickly fattens, showing a good profit on its keep. A pig with a short, pug face is not so suitable for this kind of life as one which is more capable in the jaw. Drooping ears denote a coarseness and close relationship to the old-fashioned and unimproved breeds. The neck should be wide, to give squareness to the frame. If the shoulders are wide, well filled up, and well set back, the pig will be thick through the heart. Although the flesh through the heart is not of the greatest value, as compared with that in other portions, it is found in all improved animals, whether pig, sheep, or cattle, that they thrive well, and are quicker at coming to maturity than those narrow in that part. If the legs are set on squarely, they give a squareness to the frame and admit of the full development of the sides. If coarse, they indicate an excess of bone and offal throughout the animal. The ribs should be well sprung and deep. This indicates a broad, thick side of bacon, especially where aided by a wide loin and deep, full flank. Wide hams, well let down to the hock, indicate hams of full flesh; also that the animals are carefully bred. Long, silky hair, and skin of moderate thickness, denote a good constitution and aptness to thrive, as they are opposed to the coarse, hard bristle of the unimproved animals.

Most of these features are common to other breeds, although fancy dictates a short snout and short neck to the Small White. However, as has been previously shown, this is not the typical pig for the ordinary purposes of the farm. They add to the contour of the animal, and indicate rapid maturity and smallness of offal. The Middle White should incline to the features of the Large White rather than to the Small White. The head should be shorter than in the large breed, but it should have a serviceable snout. The Berkshire should be whole-coloured, except for the blaze of white on the snout, at the tip of the tail, and on the feet. This was originally a broken-coloured pig; but experience showed that as more attention was bestowed on the breeding, the tendency was to become wholly black. It is probable that the tendency on the part of some breeders to introduce too strong an infusion of the smaller black breeds into this, with the view of toning the size, bringing about more rapid maturity, and improving the shape,—all good to a certain extent,— was fraught with danger, as rendering the breed likely to lose some of its bacon-making features, and led other breeders to see the necessity for keeping in some white. By keeping the white restricted to these two places, fixity of type has been secured. These little markings of white may appear trifling, but they answer a good purpose, and they are something more than a fancy or ornamental feature. White in other parts shows that there has been the introduction of a white breed at a comparatively recent period, and that the breed is not pure.

Telegrams—
"MOFUSSIL, LONDON."   Established 1819

A SELECTION FROM THE
PUBLICATIONS
OF
W. THACKER & CO.,
2 CREED LANE, LONDON, E.C.

AND

THACKER, SPINK, & CO.
CALCUTTA.

Demy 8vo, cloth, 12s.,

A Servant of "John Company" (The Hon. East India Company). Being the Recollections of an Indian Official. By H. G. KEENE, C.I.E., Hon. M.A., Author of "Sketches in Indian Ink," etc. With Portrait. Illustrated by W. SIMPSON, from the Author's Sketches.

"Mr. Keene has written an exceptional book. Indian biographies are often instructive, sometimes inspiring, but scarcely ever amusing. . . . Mr. Keene is not dull. This book presents a novel view of Indian life. It is the genial record of a man who from boyhood seems to have been bent on extracting the largest possible amount of pleasure from his surroundings."—*Times*.

Crown 8vo, printed on antique wove paper,
cloth extra, gilt top, 6s.,

Departmental Ditties and other Verses. By RUDYARD KIPLING. Humorous Poems of Anglo-Indian Official Life. <u>Ninth Edition.</u> With a Glossary for English Readers. Illustrated with a Frontispiece, Decorative Head and Tail Pieces, and several full-page Illustrations from Original Drawings by DUDLEY CLEAVER.

"This is the most pleasant edition of the famous 'Ditties' we have seen. Mr. Cleaver's pictures are very successful. Anyone who has not yet made the acquaintance of Mr. Kipling's early poems may be advised to do so at once."—*Literary World*.

Crown 8vo, pictorial cloth, 6s.,

Behind the Bungalow.

By E. H. AITKEN ("EHA"), Author of "Tribes on my Frontier." Illustrated by F. C. MACRAE. Fifth Edition.

"There is plenty of fun in 'Behind the Bungalow,' and more than fun for those with eyes to see. These sketches may have an educational purpose beyond that of mere amusement; they show through all their fun a keen observation of native character and a just appreciation of it."—*World*.

Crown 8vo, cloth gilt, 6s.,

The Tribes on my Frontier.

By E. H. AITKEN ("EHA"). An Indian Naturalist's Foreign Policy. Sixth Edition. With Fifty Illustrations by F. C. MACRAE.

"This is a delightful book, irresistibly funny in description and illustration, but full of genuine science too. . . . There is not a dull or uninstructive page in the whole book."—*Knowledge*.

Crown 8vo, cloth gilt, 6s.,

A Naturalist on the Prowl.

By E. H. AITKEN ("EHA"). Illustrated by a Series of Eighty Drawings by R. A. STERNDALE, F.R.G.S., F.Z.S., Author of "Mammalia of India," "Denizens of the Jungle," "Seonee," etc., who has studied and sketched animals of all kinds in their habitat and at work. Second Edition.

"It is one of the most interesting books upon Natural History that we have read for a long time. It is never dull, and yet solid information is conveyed by nearly every page."—*Daily Chronicle*.

Crown 8vo, cloth gilt, 6s.,

Lays of Ind.

By "ALIPH CHEEM" (Major WALTER YELDHAM). Comic, Satirical, and Descriptive Poems illustrative of Anglo-Indian Life. Illustrated by the AUTHOR, LIONEL INGLIS, R. A. STERNDALE, and others. Tenth Edition.

"The 'Lays' are not only Anglo-Indian in origin, but out-and-out Anglo-Indian in subject and colour. To one who knows something of life at an Indian 'station' they will be especially amusing. Their exuberant fun at the same time may well attract the attention of the ill-defined individual known as 'the general reader.'"—*Scotsman*.

Crown 8vo, cloth, 3s. 6d.,

Notes on Stable Management in India and the Colonies.

By Vety. Capt. J. A. NUNN. Second Edition, Revised and Enlarged. With a Glossary.

CONTENTS: Food, Water, Air, and Ventilation, Grooming, Gear, etc.

Captain Nunn's experience has been gained by long residence in various parts of the Indian Empire and South Africa. This little work should be in the hands of all interested in the well-being of Horses and Cattle.

Demy 8vo, cloth, 9s.,

Tactics: as applied to Schemes.

By Major J. SHERSTON, D.S.O., P.S.C., The Rifle Brigade. With an Appendix containing Solutions to some Tactical Schemes by Captain L. J. SHADWELL, P.S.C., The Suffolk Regiment, D.A.A.G. for Instruction. Second Edition, Revised and Enlarged. With Seven Maps.

" 'Tactics: as applied to Schemes,' by Major J. Sherston, D.S.O., Rifle Brigade, *is an admirable book intended to assist officers to apply drill-book principles to the solving of tactical schemes as set in the examinations for promotion.* A second edition, revised by Captain Shadwell, D.A.A.G. for Instruction, has just been published by Messrs. Thacker & Co., brought well up to date. *Major Sherston is no* doctrinaire. *He admits that problems may have more than one solution, and is at every point very helpful.*"—*Army and Navy Gazette.*

Post 8vo, cloth gilt, 7s. 6d.,

"Echoes of Old Calcutta."

By H. E. BUSTEED, C.I.E. A most interesting Series of Sketches of Calcutta Life, chiefly towards the close of the last century. Third Edition, carefully Revised and Enlarged. With additional Illustrations.

"It is a pleasure to reiterate the warm commendation of this instructive and lively volume which its appearance called forth some few years since. It would be lamentable if a book so fraught with interest to all Englishmen should be restricted to Anglo-Indian circles. A fresh instalment of letters from Warren Hastings to his wife must be noted as extremely interesting, while the papers on Sir Philip Francis, Nuncomar, and the romantic career of Mrs. Grand, who became Princess Benevento and the wife of Talleyrand, ought by now to be widely known."—*Saturday Review.*

ALFRED MANSELL & CO.

LIVE STOCK AGENTS, SHREWSBURY.

Shorthorn Dairy Cattle.

Cows and Heifers newly calved, or close at calving, and specially selected by the Firm for Dairy purposes from the best milking herds in Cumberland, Westmorland, and Yorkshire, can be supplied every week in the year in truck and half-truck loads.

Ayrshire Dairy Cattle.

Scotch-bred Cows and Heifers of this deep-milking and hardy breed, selected for the Firm by one of the best judges of Ayrshires in Scotland, are supplied in any number required at short notice.

Channel Island, Red Polled, and Kerry Dairy Cattle.

High-class Cows and Heifers are supplied on the most favourable terms.

West Highland, Aberdeen, Galloway, and Blue-Grey Cattle.

The Firm, having an extensive connection with the leading Scotch and Cumberland breeders, can execute orders promptly and satisfactorily.

Hereford Cattle.

Well-bred Bullocks and Heifers of this breed are offered for sale every Tuesday in the Shrewsbury Cattle Market. Orders from known correspondents received by nine o'clock on Tuesday morning are in time for Cattle to be selected and forwarded during the day.

Shropshire Sheep.

Shrewsbury being the centre of the district where the best flocks are to be found, and the Firm having an old-established connection with all the leading Breeders, is in a position to supply buyers with Sheep of the best type obtainable, and suitable for any purpose required. The Annual Home Sales of nearly all the most noted breeders in England are conducted by the Firm. Extensive Sales of Rams and Ewes are also held by the Firm in Shrewsbury in August and September.

Clun Forest Sheep.

Well-bred Wethers of this hardy breed produce the highest class of mutton to be found in the kingdom. The ewes are prolific and are particularly adapted for breeding early lambs for fattening purposes. Clun Forest Sheep are bred on the Shropshire hills, and can be selected and supplied from the local markets or direct from the farms where they are bred.

Pedigree Stock.

All descriptions of Pedigree Live Stock are selected and supplied on the most favourable terms.

FARMS IN HAND.

The Purchase and Sale of Stock, and the Management of Estates, and also of Farms in hand, are undertaken in any part of the United Kingdom.

STORE STOCK SALES.

EIGHT SPECIAL SALES of Store Cattle and Sheep are held by the Firm in the Shrewsbury Cattle Market in the Spring and Autumn Months; from 1000 to 1200 Cattle and a large number of Sheep are offered at each sale. All Store Cattle are Weighed before they enter the sale ring. Commissions entrusted to the Firm receive careful and prompt attention.

Buyers and Sellers of Live Stock are invited to communicate with the Firm.

Secretaries for { The Shropshire Sheep Breeders' Association.
The Shropshire and West Midland Agricultural Society.
The Shropshire Chamber of Agriculture.

Telegraphic Address: "PEDIGREE, SHREWSBURY."

INSURE YOUR HORSES AND CATTLE

WITH THE

IMPERIAL LIVE STOCK INSURANCE ASSOCIATION Ltd.

ESTABLISHED 1878.

The Oldest Company in the Kingdom confining its business to the Insurance of Live Stock.

Head Offices: 17 PALL MALL EAST, LONDON, S.W.

CARRIAGE, SADDLE, FARM, and TRADE HORSES, HUNTERS, STALLIONS, IN-FOAL MARES, and CATTLE INSURED against DEATH from ACCIDENT or DISEASE.

Claims Paid exceed £100,000.

Royal Patronage.

This Office numbers among its insurers:—

Her Majesty the Queen, H.R.H. the Prince of Wales, H.R.H. the Duchess of Albany, also the Duke of Fife, the Duke of Portland, the Duke of Westminster, Lord Belper, etc.

INCREASED BENEFITS WITHOUT EXTRA COST.

Prospectuses, Proposal Forms, and all information post free.

AGENTS WANTED.   B. S. ESSEX, MANAGER.

HAYWARD'S SHEEP DIPS.

HAYWARD'S YELLOW PASTE DIP.

Mixes with cold water; is the most improved form of Sulphur Dip. Prevents the Maggot Fly Striking, kills all Filth, cures Scab, and keeps Sheep Clean from Dipping time to Shearing. Does not make the wool harsh or dry. In Tins, 1s. each, for 20 to 25 Sheep.

HAYWARD'S GLYCERINE DIP.

In Paste and Liquid Forms. Best for Autumn and Winter. Protects the Sheep, and greatly assists the growth of the wool. In Tins and Drums.

HAYWARD'S SPECIFIC. Is a safe cure, and, if used in time, a preventive of Lung Worms, etc., infesting the Intestines of Sheep and Lambs. In Tins, 5s. each, to Drench 300 Lambs.

HAYWARD'S EAR MARKS for Sheep, Lambs, and Cattle. Are used by the Shropshire, Lincolnshire, and other Societies. Apply for samples and particulars.

ONLY ADDRESS—

TOMLINSON & HAYWARD, Mint Street Chemical Works, **LINCOLN.**

Established 1842.   *Agents in most Towns.*

PRESERVED WOOD FENCING, &c.,
For Estates, Railways, Collieries, &c.

Creosoted Fencing is neat in appearance and can be fixed by any handy labourer. It will last three times as long as Unpreserved Wood, including Larch, and needs no painting or tarring. Unlike Wire or Iron Fencing, it is not dangerous to Stock, being readily seen. During thunderstorms, Cattle and Horses have frequently been killed through being near to Iron Fences. Death has also often been caused by animals swallowing broken pieces of wire.

ARMSTRONG, ADDISON & CO.,
TIMBER IMPORTERS AND PRESERVERS,

Have supplied, during more than thirty years past, **Creosoted** and **Kyanised Timber** to Railways, Estates, and Collieries, for surface and underground purposes, and have received most satisfactory and never-failing testimony to its complete preservation, and the decided economy in using it in preference to Unpreserved Wood.

Illustrated Pamphlets and Testimonials relating to the durability of Preserved Timber and Sketches of Fencing, Paling, Gates, &c., on application.

DAY, SON, & HEWITT,

LARGEST MAKERS IN THE WORLD OF

ANIMAL MEDICINES.

Used on the Royal Farms at Windsor, Osborne, and Sandringham (by Special Warrants of Appointment), and by the Leading Agriculturists of the British Empire for over 60 years.

Will keep good for many years in any climate.

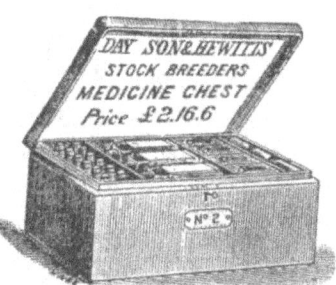

Contains every requisite for farm use.

Complete in Three Sizes, Prices: £6, 8s., £2, 16s. 6d., and £1, 8s. 9d.

For Kicks, Cuts, Bruises, Sore Withers, Swellings, Strains of Ligaments and Tendons, Saddle Galls, etc. It rapidly relieves Straining and Paining after Calving and Lambing, and is the remedy for Swollen Udders and Sore Teats.

Price 2s. 6d., 3s. 6d., and 7s. per bot.

A sure remedy for the Fret, Colic, or Gripes, Influenza, Loss of Appetite in Horses. For Debility, Scour, or Diarrhœa in Horses, Cattle, and Sheep. Hoven or Blown Cattle and Sheep instantaneously relieved.

Price 1s. 9d. per bot.; 20s. per doz.

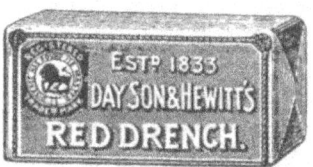

For all Feverish Disorders among Cattle, Sheep, and Pigs, and for Colds, Indigestion, Costiveness, Yellows, Surfeits, Loss of Cud, Hide-bound, Red Water, etc. Admirably adapted for cleansing and checking Feverish Symptoms in Cows and Ewes after Parturition, rendering the milk copious, pure, and wholesome. It acts as a safe preventive of Milk Fever.

Price (Cows), 13s. per doz. box; (Ewes), 3s. 6d. per doz.; 3 doz. box, 10s.

For Blood Disorders in Horses, Eczema, Surfeits, Nettle-Rash, Ringworm, Itching, Ill-Condition, Off-Appetite, Staring Coat, and Sluggishness. Invaluable for Sterility or Barrenness in Horses and Mares by virtue of its Phosphoric and other stimulating ingredients. Contains no injurious drug, and can be given at all times to all breeds of horses without stopping their work.

Price 5s. 6d. per doz. packets; 3 doz., 15s. Sold at reduced rates in tins, 10s. 6d. & £1.

ROYAL ANIMAL MEDICINE MANUFACTORY—

22 DORSET ST., LONDON, W. (Estab. 1833.)

For further Particulars regarding the "Cream Equivalent" or the "Bibby" Cake. Apply to J. Bibby & Sons, Exchange Chambers, Liverpool.

Note the Size, Age of Animal, and the Method of Rearing as described below.

Photographic reproduction of a

Pair of remarkably well-grown young Bulls—nine and eleven months old respectively—reared on "Cream Equivalent" and "Bibby" Cake. Property of W. McEwen Smith, Esq., Henbury.

Good Illustration of what may be done in the direction of early maturity.

www.ingramcontent.com/pod-product-compliance
Lightning Source LLC
Chambersburg PA
CBHW062323220526
45469CB00008B/2600